QUANTITATIVE TEXTURAL MEASUREMENTS IN IGNEOUS AND METAMORPHIC PETROLOGY

Processes involved in the development of igneous and metamorphic rocks involve some combination of crystal growth, solution, movement and deformation, which is expressed as changes in texture (microstructure). Recent advances in the quantification of aspects of crystalline rock textures, such as crystal size, shape, orientation and position, have opened new avenues of research that extend and complement the more dominant chemical and isotopic studies.

This book discusses the aspects of petrological theory necessary to understand the development of crystalline rock texture. It develops the methodological basis of quantitative textural measurements and shows how much can be achieved with limited resources. Typical applications to petrological problems are discussed for each type of measurement. The book has an associated webpage with up-to-date information on textural analysis software, both commercial and free. This book will be of great interest to all researchers and graduate students in petrology.

MICHAEL HIGGINS is Professeur des Sciences de la Terre at the Université du Québec à Chicoutimi, Canada. He is a member of the Geological Association of Canada.

QUANTITATIVE TEXTURAL MEASUREMENTS IN IGNEOUS AND METAMORPHIC PETROLOGY

MICHAEL DENIS HIGGINS

Sciences de la Terre,
Université du Québec à Chicoutimi,
Canada.

CAMBRIDGE UNIVERSITY PRESS
Cambridge, New York, Melbourne, Madrid, Cape Town, Singapore,
São Paulo, Delhi, Dubai, Tokyo

Cambridge University Press
The Edinburgh Building, Cambridge CB2 8RU, UK

Published in the United States of America by Cambridge University Press, New York

www.cambridge.org
Information on this title: www.cambridge.org/9780521135153

First published 2006
This digitally printed version 2010

A catalogue record for this publication is available from the British Library

ISBN 978-0-521-84782-7 Hardback
ISBN 978-0-521-13515-3 Paperback

To Ester and Zoe

Contents

Acknowledgements

The inspiration for this book came from a short course that I gave in September 2001 in Ankara, Turkey. I would like to thank Durmus Boztug, Cumhuriyet University, Bahadir Sahin, MTA Anakara, the Turkish Chamber of Geological Engineers, TUBITAK and MTA, Anakara for making that short course possible. This book was started while I was on sabbatical at Victoria University, Victoria, Canada and I would like to thank the department, and particularly Dante Canil for hosting me. I would like to thank my readers for their comments: Marian Holness, Cambridge University, Ned Chown, Paul Bédard and Pierre Cousineau, Université du Québec à Chicoutimi, Dougal Jerram, Durham University, Lawrence Coogan, University of Victoria, Keith Benn, University of Ottawa, Alison Rust, University of British Columbia, and Judit Ozoray. J. M. Derochette drew my attention to the Benson plate. Data were supplied by Ilya Bindeman, Margaret Mangan and Ronald Resmini. Patrick McLaren of Geosea Consulting, Victoria, showed me how sediment sizes are measured. I would also like to acknowledge Bruce Marsh, Johns Hopkins University, and Kathy Cashman, University of Oregon, without whom CSD research would never have moved along so fast. The Natural Science and Engineering Research Council of Canada and the Fondation de l'UQAC have supported my research and have enabled many of the studies presented here. Last I would like to thank my colleagues at UQAC, my students and my family.

1

Introduction

> I do not know what I may appear to the world, but to myself I seem to have been only like a boy playing on the sea-shore, and diverting myself in now and then finding a smoother pebble or a prettier shell than ordinary, whilst the great ocean of truth lay all undiscovered before me.
>
> *Sir Isaac Newton*

Newton's 'Ocean of Truth' seems to me more like a landscape: the plains are densely populated with information and most researchers work there. It is not easy to get a perspective on such a mass of information without climbing the surrounding hills. Valleys in the mountains may be difficult to find, and sparsely populated, but some lead to new basins of information ready to be explored. Other valleys may be so deep that we can glimpse what they contain, but cannot explore them closely, or even at all. This book is a guide to a country set in that landscape. Like real countries, its name varies according to who you ask, and the borders do not always follow geographic features. And like most travel writers, my description of the landscape is coloured by where I come from and what I have done.

1.1 Petrological methods

In petrology[1] we generally examine the results of natural experiments and have to tackle the inverse problem of what happened to what starting material to produce the rock that we observe. This problem is complex and does not usually have a unique solution: we must decide what the most likely starting material was and what the dominant processes were. The general approach is to choose a number of parameters and quantify them in the final rock. A starting material is proposed and various processes applied to this material. The results of this modelling are then compared to the rock in question.

All rocks comprise an assemblage of mineral grains of different sizes, orientations and shapes, which may be observed in many different ways. Early petrological studies of rocks used observations in the field and with a polarising microscope. The value of chemical analyses was appreciated, but the effort involved to analyse a small number of samples was considerable. The development of instrumental methods of chemical and isotopic analysis changed this and now such methods have become very important. In fact, many geologists view igneous and metamorphic petrology as simply applied mineral, chemical and isotopic analysis. This is a rather narrow view and in this book I hope to show that quantitative analysis of textures can also help unravel the evolution of materials. In a way, textural methods are more direct than geochemical or isotopic methods as the most important processes involve changes in the size, shape, position and orientation of the crystals. However, it is better to view all these methods as complementary.

This book is aimed at researchers in igneous and metamorphic petrology at all levels. Some workers may not be aware of the potential of textural studies to solve problems that they have previously attacked only with chemical or isotopic methods. Others may have started textural studies, but are unaware of what has been done in other subfields of petrology.

An ideal petrological study would combine relevant theory, appropriate analytical methods and application to real rocks. However, petrologists are human and in many cases one aspect is developed to the detriment of the other: the siren calls of theory and methods seem to be heard more often than application to rocks. I hope that readers will be able to combine both high-quality measurements with well-grounded theory to advance our understanding of the origin of rocks.

1.2 Qualitative versus quantitative data

Although the patterns and structures of rocks have been observed and recorded since early times, it was only in the nineteenth century that the development of the polarising microscope enabled a vast increase in the detail and number of petrological studies. At that time such observations were mostly qualitative – such as average grain-sizes, grain relationships, grain boundary shapes and orientation fabrics. While qualitative data have their uses, they cannot constrain physical models of processes in the same way that quantitative data can (for a history of quantification in geology see the review of Merriam, 2004). Models can be developed that make quantitative predications which can only be verified if accurate quantitative observations of rocks are available. All too often authors have developed elaborate mathematical models of geological processes, only to remark at the end of the article that their results qualitatively resemble actual rocks. What are clearly needed are quantitative measurements of rock textures that can be compared to theoretical predictions.

The division between qualitative and quantitative data is somewhat artificial and relates to aspects of data quality. In quantitative studies data quality refers largely to the accuracy of the measurements. If data are not accurate then they may be no more useful than qualitative data and indeed may be detrimental. In good geochemical studies accuracy is proved by analysis of geochemical reference materials. Although there are a few such materials available for textural analysis they are of limited application and are rarely used.

1.3 What do I mean by texture?

The texture of a rock is considered here to be the geometric arrangement of grains, crystals, bubbles and glass in a rock. I use the term rock here in its most general sense: for both lithified and loose materials. Materials scientists and geophysicists also use the term polycrystal to describe the same sorts of materials. Although the internal structure of grains and crystals is very important, and may help our interpretations, it is not developed here. The term texture as used here includes orientation of crystals, which is sometimes separately referred to as fabric. In other fields the terms fabric and texture are used in the opposite way. In materials science the term microstructure covers the same field (Brandon & Kaplan, 1999) and this is favoured by some geologists (Vernon, 2004). However, the name of the subject is not really important, as compared to its content.

A number of textural components can be, and are commonly quantified:

- Size of grains, crystals and bubbles.
- Shape and boundary convolutions of grains and crystals.
- Orientation of grain and crystal shapes and crystal lattices.
- Position and connectedness of crystals.
- The relationships between different phases.

Other textural components, such as enclosure relationships, are important but are not usually quantified. They can be useful as earlier textures may be preserved within oikocrysts or other structures (e.g. Higgins, 1998).

The precision and accuracy of textural parameters determined from rocks is very variable, but does not correlate with the relevance of these parameters to current problems in petrology. However, if a parameter can be readily and accurately measured, then it should be reported or made available in an electronic archive, even if its relevance is not apparent when the study is done: we have little idea what will be needed for future studies.

Quantitative textural analysis of rocks must start with measurement of textural parameters in rocks. Modelling of the evolution of these parameters is generally done by taking each parameter individually. For example crystal size is generally modelled independently from other textural parameters (e.g. the classic study of Cashman & Marsh, 1988). This probably reflects on the youthful nature of the field, as compared to geochemical or isotopic studies. However, there have been some attempts to model the whole 3-D texture of a rock mathematically.

1.4 Information density and data sources

One of the problems inherent in a book such as this is the huge variation in the number of studies in different branches of the subject. I call this the information density. For example, there are hundreds, or maybe thousands, of studies of the orientation of crystals in metamorphic rocks, yet probably less than ten in igneous rocks. This means that in some fields I have just summarised the field, referred to a specialised text or discussed a few key studies, whereas in others I mention all work in the subject. A subjective assessment of information density is shown in Table 1.1.

One of the most important roles of this book is to broaden the application of methods across the subject, and hence to even out these variations in information density.

Scientific data are circulated in many forms: some are rigorously refereed and easily accessible, whereas others do not meet these criteria. In this work I have not referenced or used short conference abstracts as they are not really refereed and rarely contain enough information to be useful. Often they seem to be just marking out territory. Many M.Sc. and Ph.D. theses do contain useful data and methods, but they are not refereed to the same standards as most journal

Table 1.1   *Estimates of information density in quantitative textural studies: O = little work, OOO = many studies.*

Field of study	Igneous petrology	Metamorphic petrology
Crystal and pore size	OO	O
Crystal shape	O	O
Crystal orientation	O	OOO
Crystal position	O	O

papers. In addition, theses are not always easy to come by and their existence and content tends to be passed on by word-of-mouth. Hence, I have generally omitted theses: high-quality studies will generally be published anyway by the student or the supervisor.

1.5 Structure of this book

Although this book covers a wide range of quantitative textural approaches to petrological problems many share similar analytical methods. Hence, I start in Chapter 2 with such generally applicable methods. I will then divide the subject on the basis of the parameters measured: size, shape, orientation and position. In Chapters 3 to 6 I will briefly review enough theory so that the nature of the problems can be understood. I then continue with a survey of the analytical methods, where they differ from the more generally applicable methods discussed in Chapter 2. The data produced by some analytical methods require significant treatment to extract meaningful textural parameters. I will present some applications that are typical of this subject, innovative, or which show the way that the subject may develop in the future. If I make too many references to my work and that of my students, then it is because I am familiar with the data, and can recalculate them to compare them with the data of others. Finally, I finish with Chapter 7 on all aspects of the textures of pores.

1.6 Software applications for quantitative textural studies

The analytical methods described in this book use a variety of different software applications. Many researchers are unwilling or unable to invest the time and money needed to acquire new software and the skills to use it. This is particularly evident if the software is not free-standing, that is, it must be used with other applications. These factors result in a tremendous range in the application of new methods that is not entirely dependent on their overall scientific usefulness. A caveat for researchers is that if you want your method to be widely employed, then you must make it easy and cheap to use. Software types and mathematical procedures are presented below in decreasing order of likelihood of use.

1. **General-purpose image processing and spreadsheet programs (e.g. Adobe Photoshop and Microsoft Excel)**. These are available for a number of operating systems (Windows, Mac, Linux, etc.). A lot of data reduction can be done with such programs, but they are commonly oriented towards business applications and hence may involve rather complex manipulations to do simple calculations.

Sometimes it is easier to augment these programs by writing a small program to do one step in the processing or to convert data formats.

2. **Special programs**: freeware, shareware and commercial. Some programs are readily available (e.g. CSDCorrections, NIHImage) while others can be very expensive. Most freeware is not available in all operating systems. However, Java programs (e.g. ImageJ) can be used on most operating systems. There are a number of large and comprehensive statistical packages that can be used to manipulate data (SAS, SPSS).

3. **Macros (procedures) for high-level languages, such as Mathematica, Maple, Matlab (Middleton, 2000), ImageJ and IDL**. Such languages are very powerful but are not so readily available. They have a steep learning curve for beginners. Many academic researchers may not realise that they already have access to these packages, sometimes via engineering departments. Many compiled IDL programs can be run, but not modified, with the aid of a free program called a 'Virtual Machine'.

4. **Uncompiled program code in common languages such as FORTRAN, C, Pascal or Basic**. Such code must be compiled for the operating systems that will be used. Many compilers are commercial and can be expensive. Minor code modifications are commonly needed and necessitate some knowledge of programming or a helpful assistant.

5. **Mathematical equations and algorithms**. Some equations are easily applied but others necessitate that software must be written before the method can be applied.

6. **'Home-made' software**. Software development with 'visual' compilers like Visual Basic, Visual C or Delphi (= visual Pascal) is relatively easy for simple applications, but difficult to perfect. If you have no programming experience you should expect to spend a week before you can make a simple program. Visual Basic is slow and should not be used for calculation intensive applications. Visual C or Delphi are not much more difficult to learn and produce a faster program.

All methods should ideally be tested with standard materials or synthetic data derived using independent methods, but this is not always available. Software 'bugs' and operating system incompatibilities are inevitable – producers of scientific software cannot usually test their programs on all systems and generally welcome constructive comments. They are generally geologists and not programmers. They do not like to hear 'your program never worked so I abandoned it'.

Software mentioned in this book is described in the Appendix and also on a website, which will be updated regularly (http://geologie.uqac.ca/%7Emhiggins/CSD.html).

Notes

1. In this book I will address igneous and metamorphic petrology. Quantification of textures is an important part of sedimentary petrology, particularly grain-size analysis of unconsolidated sediments, but the subject is too vast and distant from the petrology of crystalline rocks to be bridged by one author in one book.

2

General analytical methods

2.1 Introduction

2.1.1 Observations in three and two dimensions

Many different analytical methods have been applied to quantifying the textures of rocks, but all start with observations of rocks either in three or two dimensions. In this chapter I will start with general analytical methods that apply to the determination of many different textural parameters. Methods specific to a particular textural parameter will be discussed later in the relevant chapters. It should be remembered that different methods can be combined within a study. This can validate data acquired by innovative methods and also extend the range of grain sizes that can be quantified (Figure 2.1). Some methods that I will describe are used widely (e.g. transmitted light microscopy) whereas others are discussed here because I think that they have potential for future textural studies (e.g. Nomarski imaging). Researchers do not always choose their analytical method in the most logical way: here I want to survey the field to show what others have done. However, researchers, especially students, should never forget that a lot can be done with little equipment – my first crystal size distribution study used a microscope, camera, ruler and protractor (Higgins, 1991).

The texture of many different objects or parts of objects in a rock can be described: it is important to specify at the start what you want to examine. For instance, if a crystal is fractured then the size of the fragments or the size of the original crystal can be measured. If grains are altered at their rims then the original size of the grain or the present size of the grain can be measured.

The texture of rocks is a three-dimensional (3-D) property, hence it is most directly studied by analytical methods that view blocks of rock. However, such methods are not always applicable: samples, and grains, may be too large or small for 3-D methods; the equipment may not be available or too expensive

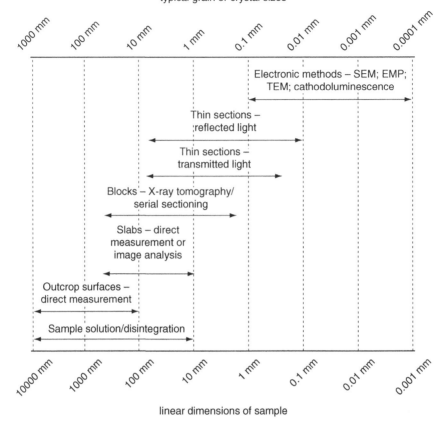

Figure 2.1   Quantitative textural methods that can be used at different scales of interest. Electronic methods include scanning electron microscopy (SEM), electron microprobe (EMP), transmission electron microscopy (TEM) and cathodoluminescence. The scale is the linear dimension of the sample – the edge of the surface or block. Maximum grain or crystal sizes that can be measured successfully are typically about one tenth of these values. Not all methods can give all textural parameters: for example, sample solution cannot give information on crystal position or orientation. Even within the scale range of a single method it may be necessary to gather the information at two different magnifications.

for the study; and many 3-D methods do not distinguish clearly touching grains of the same phase. Hence, it is commonly necessary to study sections through a rock, and extrapolate these data to three-dimensional textural parameters. The branch of mathematics that deals with this problem is known as stereology (Underwood, 1970). Detailed stereological methods will be discussed in the relevant chapters. However, here I will mention a number of *global parameters*, which describe uniquely various geometrical properties of a solid and can be determined from examination of a section. Finally, I will briefly describe some of the more general methods used to summarise the distribution of textural parameters, whether determined by 2-D or 3-D analytical methods.

2.1.2 Grain size limits

The size limits and resolution of different analytical methods must be respected (Figure 2.1). The resolution of many analytical methods is clear – for two-dimensional images it is commonly the size of the pixels (picture elements – dots or groups of different coloured dots on the screen). However, other factors may reduce the resolution, such as the lenses or physical limits. The minimum size of grain or crystal that can be measured accurately must be established from this value. For size measurements a typical limit might be ten times the resolution: for example, each measured crystal must have a length of at least 10 pixels. If a method uses discrete intervals, then there must also be sufficient grains or crystals in an interval such that the value is statistically meaningful. For example, if a size interval contains only one crystal then that value has very little meaning. These factors generally limit the maximum size of grains or crystals that can be quantified to about one tenth of the size of the image.

2.1.3 Edge effects

Almost all analytical methods, both 3-D and 2-D, must deal with the problem of edge effects of samples. Boundaries of a sample commonly intersect some crystals, which must be dealt with in a consistent manner. If the sample contains many crystals then edge effects can be ignored. For smaller samples two simple solutions are possible. The area or volume of measurement need not have a simple shape and can be a complex envelope that only includes completely visible crystals (Figure 2.2a). The area or volume of measurement can be a simple rectangle or parallelepiped that is drawn inside the actual limits of the sample so that partially visible crystals are excluded (Figure 2.2b). Then only those crystals that touch two sides of the square, or three edges of the parallelepiped, are included.

2.1.4 Textural development sequences

When a rock is sampled all that can be generally observed is the final product – the path taken to achieve that product is not always clear, but may sometimes

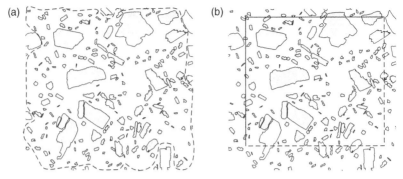

Figure 2.2   Two solutions to the problem of edge effects in textural measurements, shown here for a section. (a) Crystals that are not completely visible in the field of view (open outlines) are excluded and an irregular envelope passing midway between the crystal edges (dotted line) encloses the crystals to be measured (grey outlines). The area of the envelope is the area measured. (b) A rectangle is drawn around all crystals that are completely visible. Crystals are counted that fall within the rectangle or touch two of the sides. Those that touch the other two sides (dashed) are excluded as well as those completely outside the rectangle.

be revealed by carefully examining other, less developed samples, or by locating early textures frozen in by other processes. Such sequences of textural development can help enormously in understanding petrogenetic processes. For instance a series of lavas or samples taken from a lava lake may show how magmas crystallise (e.g. Cashman & Marsh, 1988). First-formed plutonic textures may be seen in oikocrysts and early metamorphic textures may be preserved in porphyroblasts (e.g. Higgins, 1998). Hence, the methods described below should be applied not only to the average rock, but also to special sectors of a sample that can show evidence of earlier textures.

2.2 Complete three-dimensional analytical methods

Three-dimensional analytical methods that conserve the size, shape, orientation and position of the crystals will be discussed first. Some of these methods conserve the sample (e.g. X-ray tomography) whereas others are destructive (e.g. serial sectioning).

2.2.1 Serial sectioning

The complete texture of a sample can be established from serial sectioning (Bryon *et al.*, 1995). Here, a surface or thin section is cut and recorded as a photograph or digital image. The surface is then ground away and a new surface or thin section made, parallel to the original section, and the process repeated. Clearly, the sample is either destroyed or reduced to a series of thin sections. The resolution of the method is limited by the spacing of the sections and the resolution of each image, which should ideally be equal (Marschallinger, 1998b, Marschallinger, 1998a). The images can be processed as for surface methods to separate out the different minerals (see Sections 2.6.2 and 2.6.3). The processed images can then be combined into a data volume (3-D image) to establish the complete shape of each crystal (Marschallinger, 2001).

Serial sectioning can give excellent results, especially for small numbers of irregularly shaped objects, but it is very time consuming and its resolution is limited by the spacing of the sections. For example, 500 successive images must be obtained and combined to match an image with a resolution of 500×500 pixels. This is rarely done and the vertical resolution is generally much less than the resolution in the plane of the images. If the sample is ground away ('lapped') to make the separate images then a resolution as small as 40 μm has been achieved (Marschallinger, 1998a). If thin sections are used then a much larger spacing is needed, typically several mm. Of course, the vertical resolution must be balanced by the need and ability to distinguish individual crystals. Some textural parameters do not need crystals to be separated (e.g. intercept orientation method) and in some rocks crystals can be readily isolated in plane surfaces without optical orientation. However, if crystals must be separated then it is generally easier to do in a thin section than on a flat surface.

Serial sectioning has been used more extensively in biological and palaeontological studies where the interest is in small numbers of very irregular objects – whereas petrology is more unusually concerned with large numbers of similarly shaped objects, like crystals.

2.2.2 Optical scanning and confocal microscopy

Optical scanning and confocal microscopy are special techniques that can be used for the measurement of small proportions of grains in transparent materials. They give a result similar to serial sectioning, but without destroying the sample. In optical scanning the section is examined at high magnification with a large aperture. In this situation the depth of field is small and a narrow range of depths in the section are focused. A photograph is taken and the sample–objective distance increased. The process is repeated to build up a complete 3-D reconstruction of the section. The matrix of the crystals must be sufficiently transparent and the crystal number density must be sufficiently low that the whole crystal can be observed; hence it can only be applied in special circumstances, for instance microlites in a glassy volcanic rock (Castro *et al.*, 2003).

A confocal microscope is designed specifically for these types of application (e.g. Petford *et al.*, 2001, Bozhilov *et al.*, 2003). Instead of shining a light on the whole section, only a part of the sample is illuminated with a laser beam that is scanned across the sample. The image is then reconstructed sequentially, as in a scanning electron microscope (see below). This method has the advantage that scattering of light from adjacent crystals and matrix into the volume of interest is much reduced.

It is not always necessary to reconstruct the whole 3-D structure: the length and other shape parameters can be determined from the vertical and horizontal position of the ends of the crystal. In some cases the method has been simplified further by choosing crystals that are nearly parallel to the plane of the section (Castro *et al.*, 2003).

2.2.3 X-ray tomography

Tomography (CAT – computed axial tomography) is the reconstruction of a section from many separate projections around an object. A series of closely spaced slices are assembled into a 3-D image. It is commonly applied using X-rays (Ketcham & Carlson, 2001, Mees *et al.*, 2003), but can also be used with any radiation that can penetrate the material, such as gamma rays or even light (Figure 2.3).

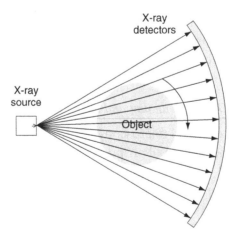

Figure 2.3   X-ray tomography. In most systems used for geological research the sample is rotated as it is bathed in a flat beam of X-rays. The X-rays that pass through the sample are detected by a curved bank of detectors. A slice of the internal structure of the sample can be reconstructed from the quantity of X-rays received by the detectors. The sample is then moved vertically and another slice analysed.

Minerals are distinguished in X-ray tomography on the basis of their linear attenuation coefficient, μ. This depends directly on the density of the mineral, the effective atomic number of the mineral, and the energy of the incoming X-ray beam. Common minerals have values of μ that vary by a factor of three for 100 keV X-rays, which is much greater than the precision of measurement, typically 0.1%. However, there is much overlap in the values of μ, hence minerals cannot always be distinguished using this parameter alone. In some situations the sample can be examined using X-rays of two different energies (frequencies) and the images combined to separate phases. In general, this method cannot separate touching crystals of the same mineral. This is because μ does not depend on direction in an anisotropic mineral. This contrasts with the optical properties of most common anisotropic minerals.

Sample size is limited by the attenuation of X-rays and the physical dimensions of the sample chamber. Many instruments can handle samples up to 40 cm long. Spatial resolution is a function of sample size and the number of pixels in the image (typically less than 1000×1000 pixels). Resolutions are commonly 0.1–0.2 mm for centimetric to decimetric size samples. Larger samples necessarily have a lower resolution.

Recently, synchrotron radiation has been used for micro-tomography (Song *et al.*, 2001, Cloetens *et al.*, 2002, Ikeda *et al.*, 2004). This has a wide range of wavelengths and is very intense so that it can be focused in a small volume. It has also been combined with X-ray fluorescence computed tomography to give a 3-D compositional map (Lemelle *et al.*, 2004). Although precision is low, this method shows promise for some materials.

2.2.4 Magnetic resonance imaging

Another 3-D method is magnetic resonance imaging. This is used extensively in medical applications, but has only recently been applied to textural studies of rocks. Magnetic resonance generates images that largely reflect the hydrogen content of geological materials. So far this method has only been used in a single textural study, where it was used to help visualise pore distributions in carbonate rocks (Gingras *et al.*, 2002). Clearly, it may be applied in other studies of water-bearing rocks, especially if the resolution of the method can be improved.

2.3 Extraction of grain parameters from data volumes

The digital 3-D analytical methods described above produce a 3-D image commonly referred to as a data volume, or data brick. It is comprised of voxels, the volumetric equivalent of pixels in images. Two-dimensional image analysis (see Section 2.6.3) is much more developed than analysis of three-dimensional images. Many of the 2-D techniques can be extended to 3-D, but the software for this is still in development.

Extraction of grain and textural parameters from data volumes comprises three steps: classification, separation and measurement (Ketcham, 2005). The classification process is commonly the most complex. Many data volumes are essentially grey-scale images – that is there is only one value for each voxel. Such images can be segmented by considering a window of acceptable values. However, mineral phases are not always very regular, and more useful methods have been developed (Ketcham, 2005). For instance, the 'seeded threshold' filter initially accepts voxels within a range of grey-scale values. Each seed object is then expanded by the addition of connected voxels that have a wider range of grey-scale values. If the mineral phase is assumed to be spherical then irregular groups of voxels may be simplified by substituting spheres with volumes equal to that of the original crystal (Carlson *et al.*, 1995).

Touching crystals or grains are not separated by most 3-D analytical methods; hence this must be done during data reduction. Voxel groups can be examined individually and cut apart manually (Ketcham, 2005). An automatic process was suggested by Proussevitch and Sahagian (2001); interconnected voxel clusters are 'peeled' or eroded until the individual objects are separated, and finally the crystal centres are defined. The crystals are then rebuilt using an assumed shape, such as a sphere or equidimensional polyhedron. Another approach is to use the 'watershed' algorithm: the acceptable range of voxels in a group is reduced until the group separates into distinct objects. The voxel group is then rebuilt from these centres (Ketcham, 2005).

Measurement of the dimensions of separated groups of voxels is conceptually easy, but has not been facilitated by current software. However, new developments may ease this problem (Ketcham, 2005).

2.4 Destructive partial analytical methods

Some aspects of rock textures can be determined by dismantling the rock and measuring the dimensions of the separated grains. This avoids problems of interpreting the grain parameters from sections, but the position and orientation of the grains are lost. In addition, grains with convoluted shapes may not be easily separated intact from their matrix and it is difficult to know how to deal with edge effects. Such analytical methods enables the use of much smaller samples as many more crystals are encountered in volume compared to those intersected in a section. However, the minimum crystal size that can be consistently recognised must be clearly established (and also the maximum size in rare cases).

2.4.1 Sample disaggregation

Grains can be separated from a rock if it can be readily disaggregated. In some volcanoes explosive eruption processes may separate crystals mechanically. However, it is unlikely that this process will be truly unbiased in terms of size or shape. Other volcanic rocks may be so weak that they can be separated mechanically with little damage to the crystals (also see sample solution methods below). Dunbar *et al.*, (1994) did this for bombs from Mt Erebus, Antarctica, with good results. However, they did have to make a correction for some broken crystals.

Well-lithified rocks can be disaggregated using more energetic methods. In electric pulse disintegration a voltage of 100 kV is applied to a rock sample sitting in a water bath (Rudashevsky *et al.*, 1995). The rock explodes, separating crystals along grain boundaries. The specialised nature of the equipment has limited the use of this method so far. Finally, it is possible to dissect a rock crystal by crystal using a hammer and chisel. Kretz (1993) used this method to measure the position of each garnet crystal in a schist. He then reconstructed the whole structure with balls and rods.

Crystals that are already fractured in the rock present special problems. If one is interested in primary processes then one solution is to examine each crystal individually and only retain those that are unbroken. This is time consuming and may bias the sampling of the crystals, but it may enable some studies that would otherwise be difficult (Gualda *et al.*, 2004). Of course, if the focus of the study is on the fracturing process then the crystals can be measured easily.

2.4.2 Sample dissolution

Dissolution is another method for the separation of crystals from a rock. In carbonate matrix rocks (carbonatites and marbles) some crystals can be separated by dissolution of the matrix using HCl or other acids. However, it should be remembered that some silicates are also slightly soluble in HCl (e.g. anorthite, olivine) and hence small crystals may be lost.

Similar dissolution methods are commonly used to separate diamonds from kimberlite or other rocks for exploration purposes. The preferred method is crushing followed by fusion with alkali flux at 550 °C. Commercially very large samples are processed – up to 300 kg. Larger crystals may be broken during crushing that precedes dissolution. In that case the crush size is the maximum crystal size that can be recognised. In some kimberlites most diamonds are already broken, probably during emplacement; hence the size measured is that of the fragments, not the original crystals. The smallest crystals may be lost by solution or mechanically.

Dissolution methods are also used to extract crystals from silicic volcanic rocks (Bindeman, 2003). The method works best for light, frothy pumice. Hydrofluoric (HF), fluorosilicic (H_2SiF_6) or fluoroboric (HBF_4) acid is used to attack the glass. The acid is applied either until there is complete dissolution of glass or until the glass is weakened by partial dissolution and the rock can be easily crushed. Many silicate minerals are also attacked by the acid, but generally much more slowly than the glass. The surfaces of feldspar crystals are etched more than quartz, which can be used to distinguish these minerals. Small crystals may dissolve, stick to the surfaces of the preparation equipment or be retained on filters. However, good results have been obtained for zircon and quartz in rhyolitic pumice (Bindeman, 2003).

Mixtures of different minerals extracted by dissolution can be separated by density using heavy liquids (methylene iodide = diiodomethane, bromoform = tribromomethane, sodium polytungstate) or magnetically using a Frantz© isodynamic separator (Hutchison, 1974). They can also be hand picked dry, under alcohol (to reduce reflections) or in immersion oil. In the latter case the refractive index of the oil can be matched to that of the matrix glass or another mineral, to make it less visible.

2.5 Surface and section analytical methods

2.5.1 Surface preparation techniques and artefacts

Most quantitative textural studies of rocks start with the preparation of an artificial flat surface: natural fracture surfaces cannot generally give quantitative results. Commonly, the first step is sawing a rock sample with a diamond-impregnated circular or wire saw. The rough surface can then be flattened on a lap (rotating wheel) with abrasive paste. The surface is polished with progressively finer grained abrasives until the necessary degree of flatness has been achieved. A number of problems and artefacts are commonly encountered: they should be recognised so that they will not be misinterpreted. This is especially a problem with automatic analysis systems.

- **Scratches**: These are not always easy to remove. They can be a problem if the rock is comprised of minerals with variable hardness. Surface treatments like etching can enhance small scratches.

- **Occluded materials**: Grains of the grinding material can be pressed into softer minerals, or can be caught along grain boundaries or cracks in the sample.
- **Pull-outs**: Brittle minerals with well-developed cleavages can fracture close to the surface making small pits.
- **Rounding and surface relief**: In polymineralic rocks harder minerals will resist abrasion and will tend to be higher in the final polished surface.

The relief of the surface can be increased by etching. This is commonly used for Nomarski microscopy (see Section 2.5.3.3) but has also been applied in other studies. Herwegh (2000) developed a two-stage etching for calcite: the surface is first immersed in dilute HCl, followed by dilute acetic acid. The surface relief was then examined with a scanning electron microscope; however, Nomarski microscopy could also be applied.

2.5.2 Electronic and associated analytical methods

Rock surfaces can be examined using a beam of electrons and the most common instrument for this is the scanning electron microscope (SEM; Reed, 1996). The sample is placed in a vacuum chamber and a beam of electrons is scanned across the surface. Interaction of the beam with the material produces electrons, X-rays and light photons that are detected and measured. The resolution of the different images is variable, but is theoretically smaller than for optical measurements as the wavelength of electrons is much smaller than that of light.

Samples must be flat and highly polished otherwise it is the relief that will be imaged instead of the composition. Solid samples or polished thin sections can be used, with the only physical limitation being the size of the sample chamber, typically less than 10 cm. Samples are commonly coated with carbon or metal to make them conducting and to reduce charging of the surface (Reed, 1996). In some instruments the sample chamber is kept at a higher pressure than the electron gun and hence no coating is necessary (ESEM – environmental scanning electron microscope). The magnification can be varied enormously, making this technique very useful over a wide range of sample sizes from 0.1 μm to 1 mm.

The electron microprobe (EMP) is another instrument that uses electrons to analyse materials. It is mechanically very similar to the SEM, but has been optimised for different measurements. SEMs are designed for observations at different magnification scales, but the sample cannot be viewed optically. EMPs commonly have a magnification fixed to that of the associated optical system. In addition EMPs are optimised for quantitative chemical analysis. Recently, there has been a convergence between these two instruments, but SEMs are still cheaper than EMPs for both purchase and use.

2.5.2.1 Backscattered electron images

When a beam of electrons strikes a surface some electrons will pass close to the nucleus of the atoms and will be scattered by the positive charge of the protons. If the electron beam is approximately normal to the mineral surface ('flat scanning') then the number of 'backscattered electrons' (BSE) emitted by any part of a rock is proportional to the mean atomic number of the mineral, $\bar{Z}$:

$$\bar{Z} = \frac{\sum Z_j N_j R_j}{\sum Z_j R_j}$$

where Z_j = atomic number; N_j = atomic weight and R_j number of atoms in the formula of element j. If the beam is strongly inclined to the surface then other factors come into play (see orientation contrast imaging below).

The spatial resolution of BSE images is limited to about 0.1 μm as the electrons are produced within a relatively large volume. The atomic number difference that can be distinguished in a BSE image also decreases with increasing atomic number (Reed, 1996): At $\bar{Z} = 10$ u (e.g. quartz, feldspars, Table 2.1) it is 0.1 u and at $\bar{Z} = 30$ u (e.g. Cu-sulphides) it is 0.5 u. However, the overlap of $\bar{Z}$ ranges of potassium feldspar, plagioclase and quartz is much more of a problem, hence supplemental information, such as X-ray maps may also be necessary. BSE images are most useful for small crystals that have significantly different $\bar{Z}$ from other minerals and the groundmass.

Table 2.1 *Mean atomic numbers ($\bar{Z}$) for various minerals. These are derived from actual analyses in the MinIdent-Win 3 mineral property database (see Appendix).*

Mineral	Mean atomic number ($\bar{Z}$)
quartz	10.8
albite	10.8
anorthite	11.9
orthoclase	11.7
fayalite	18.3
forsterite	11.4
orthopyroxene	13.8

2.5.2.2 X-ray maps: EMP, SEM and Micro-XRF

X-ray maps are images in which the intensity (pixel value) is related to the composition of the surface. They are created from analysis of secondary X-rays produced when electrons strike a surface with sufficient energy (Reed, 1996). The energy (or wavelength) of some of these X-rays are characteristic of the atomic number of the atoms in the target and hence can be used to determine the elemental concentration. The quantity of X-rays produced per unit mass decreases considerably with atomic number, as does the sensitivity of the detector and window systems.

All EMPs and most SEMs have the X-ray detectors needed to produce X-ray maps. There are two types of X-ray detector. Energy-dispersive detectors use a silicon or germanium crystal and can measure many different X-ray energies simultaneously. Most are not sensitive to elements lighter than sodium and the resolution is not always sufficient to separate adjacent peaks produced by different elements. Wavelength-dispersive detectors use a crystal to diffract the X-rays produced by the sample. The angle between the detector, sample and the crystal is varied to select different X-ray wavelengths and hence only one element can be measured at a time. The resolution and sensitivity of this detector are greater than those of the energy-dispersive detectors and hence they are more sensitive and precise.

If an electron beam is scanned (rastered) across a sample then the X-rays emitted at each point can be filtered for each element. This can be assembled to give an X-ray map of the sample for each element. For larger samples the beam can remain stationary and the sample driven mechanically to give the same effect. Such images overcome some of the problems associated with BSE images, in that minerals such as quartz and feldspars are clearly distinguished. However, production of X-ray maps is time-consuming and hence costly. In addition, the resolution of the images is commonly poor compared to BSE images as the X-rays are much less intense.

Secondary X-rays are also generated when a beam of primary X-rays strikes a surface. This is called X-ray fluorescence (XRF) and is commonly used for the chemical analysis of powder samples. Recent developments in X-ray optics have enabled the beam size to be reduced to 50 or even 10 µm in a micro-XRF instrument. While this resolution is much less than that of an SEM it is sufficient for many studies. The X-rays are generally analysed with an energy-dispersive detector, as for an SEM. There are many advantages to this technique: the equipment is much cheaper and the analyses faster. It can operate without a vacuum and there are no electronic charging effects, as X-ray photons are neutral.

2.5.2.3 Cathodoluminescence

Cathodoluminescence is the emission of light by a crystal in response to electron bombardment (Pagel, 2000). This effect is only seen in some minerals, but these include such common species as feldspars, calcite, zircon and quartz. The intensity and colour of the light are highly variable and commonly depend on the concentration of trace elements called activators and the density of lattice defects. Hence, cathodoluminescence can be used to distinguish grains or parts of grains with different growth histories. It is used extensively to examine the petrology of sedimentary rocks, but has been less exploited in textural studies of metamorphic and igneous rocks (e.g. Titkov *et al.*, 2002).

Cathodoluminescence is most commonly measured with a luminoscope: a special instrument that is attached to a regular petrographic microscope. It can also be observed with the optical system of an EMP. The most sensitive method is to use a special light detector in an SEM, but this only records the intensity and not the colour of the light.

2.5.2.4 Orientation contrast imaging

Under normal 'flat scanning' mode the mean atomic number of the crystal ($\bar{Z}$) controls most of the variation in BSE intensity. However, if the electron beam hits the sample obliquely then the crystallographic orientation of the crystal becomes important because electrons are channelled into and out of the crystals along lattice planes (Figure 2.4; Prior *et al.*, 1999). If these are parallel to the electron beam and/or direction of the detector the electron intensity will be enhanced. This effect is exploited in orientation contrast imaging (OC). In flat scanning the OC effect is about ten times less important than $\bar{Z}$ effects. However, if the sample is tilted at 70° to the beam direction then OC dominates (Figure 2.4).

Surface preparation is particularly important in OC. Normal polishing (sufficient for BSE images) disturbs the crystal lattice near to the surface and inhibits the OC effect. Therefore, final polishing must be done using special techniques, such as colloidal silica (Xie *et al.*, 2003), chemical–mechanical polishing or etching (Prior *et al.*, 1999).

This method may be especially useful for imaging touching cubic (isometric) crystals. Such minerals are optically isotropic and hence cannot be separated in reflected and transmitted light. However, OC varies with the orientation of the lattice planes; hence the crystals will have different intensities and grain boundaries can be easily identified.

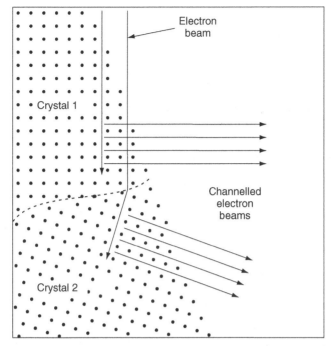

Figure 2.4 Orientation contrast imaging. The lattices of two crystals with different orientations are indicated by the arrays of points. In this example the electron beam is vertical and the sample surface is at an angle of about 70 degrees. Electrons are channelled into and out of the crystals along lattice planes.

2.5.3 Optical analytical methods

Examination of rocks in transmitted and reflected light has a long history and still remains a very cost-effective analytical method. Indeed, it is probably the most commonly used method in quantitative textural analysis. The wide range in the optical properties of minerals, especially their common anisotropy, makes it easy to distinguish individual crystals, even when they touch. Hence, it is the method of choice for minerals that are especially abundant in a rock.

The quality of images is very important for successful quantification. Many older microscopes have high-quality dedicated (and commonly antiquated) film cameras. Individual photographs taken on 35 mm film can sometimes be enlarged to 35 cm without loss of resolution. The prints can then be scanned for digital processing. Small 'domestic' digital cameras commonly have complex zoom lenses which can have considerable flare and chromatic aberration. More expensive dedicated digital cameras have much simpler and better lenses and greater resolution.

2.5.3.1 Transmitted light

Rocks are examined in transmitted light using thin sections. A standard thin section is 30 μm thick and measures 2×3 cm. Larger thin sections, 3×7 cm, are also readily available, as are thinner-than-normal sections. Polished thin sections that can also be used in reflected light are generally only available in the smaller size. Identification of minerals by optical methods is complex and is covered by basic texts, such as Nesse (1986).

Thin sections are usually examined in plane-polarised light and under crossed polarisers with a petrographic microscope. Normal petrographic microscopes equipped with film or digital cameras can yield good photographs of regions up to 2 mm. Unusual low-power objectives ($\times1$) can image larger areas, but special stages are needed and illumination is commonly uneven. Larger images can be assembled by taping together photographic prints into a mosaic or by electronically stitching digital images (see Appendix).

It is difficult to get an image of a whole thin section with a regular petrographic microscope. Some specialised zoom microscopes can cover a whole regular thin section with as few as 10 images. Other solutions found are a film scanner (Tarquini & Armienti, 2001), a slide-copying apparatus or camera attachment, or a light bench. Some systems use a flash so that vibration problems are reduced. It is also possible to use a document scanner that can operate with transmitted light on a thin section. This works well for a direct image or with plane-polarised light produced by a single Polaroid sheet. However, if two Polaroid sheets are sandwiched on either side of the section to make a cross-polarised image then there is not always enough light to yield a good image.

The lower limit of resolution is commonly close to the thickness of the section, 30 μm. Although the physics of light indicates that a higher resolution is possible, it is rarely approached in most studies because of the difficulty of observing small crystals in thin section, except where the matrix is lightly coloured. It should also be remembered that small crystals enclosed in the thin section are seen in projection whereas larger crystals are seen in section. This necessitates different mathematical procedures to calculate the 3-D textural parameters (see Chapter 3).

A multitude of colour images are possible from a single thin section: in plane-polarised light pleochroic minerals will have different colours according to the orientations of the polarisers. Similarly, when examined with crossed polarisers crystals will show different colours and extinctions according to the orientation of the polarisers. Most types of automatic image analysis cannot cope with more than a single image, however, computer-integrated polarisation microscopy methods are designed to exploit this aspect of optical mineralogy (see Section 2.6.4; Heilbronner & Pauli, 1993, Fueten, 1997).

In cross-polarised light some crystals will generally be in extinction for any orientation of the section. This can be a problem for the measurement of crystals using simple observations of a single cross-polarised image. A little-used optical technique, the 'Benford plate' can eliminate this problem (Craig, 1961). The name is a misnomer as the set-up consists of two matched quarter wave plates (retardation = 132 nm). One is placed below the thin section, oriented at 90° to an accessory plate holder. The other is inserted in the normal position for an accessory plate. With the Benford plates installed, the birefringent colour of crystals is identical to that seen at the 45° position without the plates. However, the birefringence colour, and its intensity, do not change with rotation of the stage; that is, extinction no longer occurs. Clearly, this can be very useful as all crystals can be observed with the stage in a single orientation. However, the intensity contrast between grains may be less than that normally observed. This image is identical to the maximum birefringent image obtained in computer-integrated polarisation microscopy (see Section 2.6.4).

Thin sections are generally observed orthogonal to the plane of the section, but in some situations it may be useful to remove this constraint. The universal stage is an accessory for a standard petrographic microscope that enables tilting of the section with respect to the direction of observation. Hence, crystals and their boundaries can be examined from different orientations. It is described in more detail in Section 5.5.1.

The infrared absorption of mineral sections can also be examined, generally by using a high resolution infrared spectrometer (e.g. Ihinger & Zink, 2000). A section thicker than normal (800 μm) is polished on both sides and the spectra of 100 μm spots collected. Each spectrum is filtered for absorbance in a range of wavelengths that correspond to resonance of specific chemical entities and is used to build an absorbance map of the section. This technique has been used to examine the distribution of hydrogen in quartz.

2.5.3.2 Reflected light

Rocks are examined in reflected light using polished thin sections or blocks (Cabri & Vaughan, 1998). Examination of polished sections in orthogonally reflected light can complement studies in transmitted light or be done independently. Many research microscopes are set up so that they can be used for both optical techniques. The variation in optical properties in reflected light is generally more limited than that in transmitted light, but the method is especially important for opaque minerals. The resolution of the method is not limited by scattering of light within the material and hence features as small as

1 µm can be readily observed. However, there is no good way to get images of areas larger than can be viewed with the normal microscope lenses or assembled from image mosaics. For instance, a document scanner does not yield very good images because the light is not orthogonal to the surface. However, it may be useful if there are very large differences in reflectivity.

Although surfaces may be viewed in both plane-polarised light and under crossed polarisers, the former is used most commonly for imaging surfaces. As for transmitted light, anisotropic minerals may have different reflectances (colours) for different orientations, but the effect is not large for most minerals. Hence, minerals are usually imaged in plane-polarised or unpolarised light. In this situation the most important parameter is the reflectance of the mineral. Reflectances have been recently tabulated for all new minerals, at 20 nm intervals from 400 to 700 nm, but values at four standard wavelengths are generally used (470, 546, 589 and 650 nm). Most transparent minerals, such as quartz and feldspars have reflectances in the range 5–10%. Other minerals are much higher: oxide minerals 12–30%; sulphides 12–60% and metals 50–100%. Surfaces may also be etched or stained to bring out the contrast between different minerals, sub-grain boundaries and crystal defects (see Section 2.5.3.3 and review by Wegner & Christie, 1985).

Minerals can be imaged electronically by standard charge-coupled device (CCD) cameras. However, such cameras are designed to mimic the human eye and hence record light in three wide spectral bands. A better approach is to use narrow (10 nm) bandwidth filters (Pirard, 2004). Such a system needs to be carefully calibrated, but can give much better resolution of mineral phases, especially those with reflectances greater than 5%. This method can distinguish mineral pairs that are problematic in BSE (e.g. chalcopyrite/pentlandite) or X-ray maps alone (e.g. hematite/magnetite/goethite).

2.5.3.3 Nomarski (DIC) microscopy

A very useful optical technique for imaging surfaces is differential interference contrast (DIC) or Nomarski microscopy. It has long been used by metallurgists to produce detailed images of polished surfaces, and has also been used by petrologists to examine mineral and rock textures (Anderson, 1983, Pearce *et al.*, 1987, Pearce & Clark, 1989). It can reveal both the exterior shape of crystals and their internal structure, and can be useful for distinguishing crystals from a glassy matrix. It can be applied to thin sections and hence is complementary to normal reflected and transmitted light methods.

Nomarski microscopy is an optical technique for imaging the micro-relief of surfaces (<0.5 µm). A light beam is split by a prism; one beam is reflected off the surface and allowed to interfere with the reference beam. Hence the method

Table 2.2 *Etchants used for Nomarski examination of minerals. All samples should be neutralised with Na_2CO_3 after etching so that degassing of HF does not etch the objective lenses. A much longer list of possible etchants and further details of the methods are available in Wegner and Christie (1985).*

Mineral	Etchant	Notes and reference
Plagioclase	Fluoroboric acid (HBF_4); 2–3 minutes at room temperature	(Anderson, 1983)
Olivine	Concentrated HCl; 10–20 minutes at 45 °C	(Clark *et al.*, 1986)
Clinopyroxene	Concentrated HF; 2–4 minutes at room temperature	HF attacks plagioclase and olivine vigorously (Clark *et al.*, 1986)

is sensitive to surface relief of the order of the wavelength of light. The special lenses and prisms that are needed are usually fitted to a specialised microscope, which may be available in metallurgy laboratories.

The initial surface must be very well polished, with no remaining scratches. This can be done in the same way as for normal polished sections. The surface is then etched to develop relief (Table 2.2; Wegner & Christie, 1985). Etch depth will depend on the nature of the mineral, the orientation of the section and the density of crystalline defects. Finally, the surface may be coated with carbon or metal using the same apparatus that is used for SEM studies. This coating reduces interference from reflections beneath the surface of the sample. However, it also reduces the contrast in reflectivity between different minerals and glass. If polished thin sections are used then the technique may be complemented by examination with transmitted light, if the coating is not too thick. This is useful if two anisotropic crystals touch; they may then be distinguished by differences in optical orientation.

2.5.4 Slabs and outcrops

For rocks containing large crystals sawn slabs can be very useful. The surface should be planar, but it is not usually necessary to have a perfectly polished surface, unless very small crystals will be examined. A sawn surface can be ground flat with wet abrasives on a rotating lap (wheel) or with waterproof (wet and dry) silicon carbide/oxide paper on a flat surface such as a sheet of glass.

The surface can be etched or stained to enhance the contrast between the different minerals. One of the best known stains is sodium cobaltinitrite which

Table 2.3 *Some staining procedures that can be used to help distinguish minerals in slabs. Some of these techniques can also be used for staining thin sections. Details of these and other procedures can be found in Hutchison (1974).*

Minerals	Treatment	Colours and notes
K-feldspar	1) HF (49%) 30 seconds 2) $Na_3Co(NO_2)_6$ Sodium cobaltinitrite (freshly prepared) 10 seconds	K-feldspar = orange Plagioclase = white Quartz = grey Quite resistant to abrasion
Plagioclase	Amaranth red	Red. Easily removed from surface
Calcite and dolomite	1) HCl (1.5%) 10 seconds 2) Alizarin red 3) Potassium ferrocyanide $K_4Fe(CN)_6 \cdot 3H_2O$	Calcite = rose to red Dolomite = colourless

colours all potassium feldspar a deep orange. Other minerals can also be stained (Table 2.3; Hutchison, 1974). It is also possible to stain thin sections.

Crystals in slabs can be measured manually in several ways: (1) The simplest is with a ruler and protractor (see below). (2) Mineral outlines can be traced onto a transparent overlay and subsequently scanned and analysed automatically. (3) Slabs can be placed on a document scanner and the images analysed automatically or manually. Very large polished rock panels, as used for building facing, can provide material transitional in scale between slabs and outcrops.

Very large crystals must be examined in outcrop, especially those with low number densities. Crystals can be measured directly with a ruler and protractor. Photographs of outcrops can also be used, but not generally so successfully – what is clear in the field can be confusing on a photograph at a later date. The method is most easily applied to flat surfaces naturally polished by glacial action (Higgins, 1999). However, it can also be used on surfaces smoothed by rivers or the sea. In some cases dimensional stone quarries may yield sufficiently smooth surfaces.

2.6 Extraction of textural parameters from images

Some of the analytical methods described above yield quantitative textural data directly. However, most produce images that must be measured to extract various textural parameters. Manual image analysis methods demand a large amount of individual attention and judgement and user training is very important. Such methods generally yield high-quality data directly, but are very time-consuming. In addition, there is always the possibility of operator bias, especially where the researcher does the measurement. Automatic image analysis methods may take much time to set up, but can be faster if many, similar samples need to be processed (see the general review in Russ, 1999). The quality of the data is commonly not as high as that produced manually, even though it may suffer less from operator bias. This may be balanced by a greater number of crystals and samples measured.

The images produced by many analytical methods have too few pixels for adequate precision. Several images can be combined together like a mosaic to produce a larger image. This can be done electronically or by taping photographic prints together and scanning the mosaic.

2.6.1 Size limits of measurements

No matter what method is used to acquire quantitative textural data, the minimum size of crystal that can be recognised and measured must be established. If no data are listed for crystals smaller than a certain size it is important to indicate if this reflects an artefact of measurement or a real lack of crystals in that size range. Methods can be combined to cover a wider size range than would be possible with a single method. For instance data from outcrop measurements can be combined with data from slabs; thin sections can be measured at two different scales; BSE images can be combined with thin section measurements. It is very important to record and publish the details of the data-acquisition method and steps taken to ensure adequate quality control.

The maximum size of crystal that can be precisely determined is generally determined by the number of crystals in the largest interval, and not by the physical limitations of the method. However, the maximum size must also be specified where possible.

2.6.2 Manual image analysis

Crystals viewed with an optical microscope can be measured directly without recording an image: the scale in the microscope eyepiece can be used to measure intersection length and width, and the crystal rotated on the stage to determine the orientation. This has the advantage that different magnifications and orientations of the stage can be used to identify the crystal; hence a wide range of crystal sizes can be measured. It is also useful for rocks that have a low number of crystals per unit volume (number density). However, it is difficult to keep track of which crystals have been measured, and the use of a photograph is advised, even if only to check off that a crystal has been measured. A similar method can be used with an electron microscope (SEM), with the same caveats.

Most of the techniques discussed above produce images: single images or mosaics, on paper or in electronic form. These images can then be measured manually in several different ways: (1) the intersection length, intersection width, long-axis orientation and centroid position can be determined using a ruler and protractor; (2) the crystal outlines may be traced onto a transparent overlay, which is then scanned and analysed automatically as discussed below; (3) digital images can be imported into a vector drawing program (e.g. CorelDraw^{TM}) and the crystal outlines traced by drawing a polygon with the mouse; (4) photographs can be put on a digitising tablet and the outline of each crystal in the photograph traced using the digitising puck. A number of programs can be used to control the digitising tablet (see Appendix) and reduce the raw positional data to textural parameters. In each case the actual outline of the crystal in the thin section should be verified with a polarising microscope.

2.6.3 *Automatic image analysis*

Automatic image analysis commonly involves many different steps (Figure 2.5). The goal of image processing is to segment the original image into a classified image of the mineral of interest: a wide range in pixel values in one or many channels of the original data image are reduced to a small number of different pixel values, one for each phase of interest. Clearly, it may not be easy to classify some pixels. This commonly occurs where a pixel traverses the contact between two minerals. Sub-pixel grains and unknown minerals can also cause problems. It must be decided if it is really necessary to classify all pixels or if unclassified pixels can be accepted.

A number of successful image processing procedures have been published but many are specific to particular proprietary image processing software (e.g. Perring *et al.*, 2004). The expense and limited distribution of these systems has led many researchers to use general purpose image software (i.e. Adobe Photoshop^{TM}) and powerful open-source, free software (e.g. NIHImage/ImageJ). Here, I have tried to present an approach that can be applied more generally. I do not want to enter into the specifics of program usage, as this will be outmoded before the book is published.

2.6.3.1 *Grey-scale images*

Some analytical methods produce images with a single variable, usually called grey-scale images: for instance, the amount of backscattered electrons or the

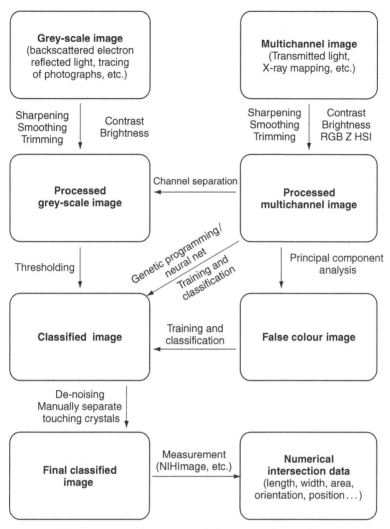

Figure 2.5   Flow diagram for automatic image analysis of grey-scale (single channel) or multichannel images. This diagram does not show all possible strategies, just the most common routes.

overall reflectance in reflected light. Grey-scale images can be smoothed, sharpened and trimmed to improve recognition of individual mineral grains. Such manipulations can be done using readily available general purpose, image-processing software (GPIP; see Appendix; e.g. Adobe Photoshop™ or Corel Photopaint™) but dedicated programs commonly make it clearer what has been done to the data and how it can be repeated (e.g. LViewPro™).

Processed grey-scale images are easily segmented by selecting upper and lower thresholds (tones of grey) for each mineral by 'training': the values associated with a mineral are determined for a known crystal. Pixels that differ considerably from those of the known minerals are best ignored, even if this means that part of the image will be unclassified.

2.6.3.2 *Multichannel images*

Many analytical methods produce images that have more than one value associated with each pixel. The image associated with each variable is called a channel. For instance, optical images of minerals may be coloured, which is generally expressed using the red–green–blue model that mimics the behaviour of the human eye. These data comprise three channels. Another approach is to use narrow (10 nm) bandwidth filters, which can yield many more colour channels (Pirard, 2004). X-ray images of surfaces can have a channel associated with each element. Such multichannel data avail themselves to more complex processing than grey-scale data. However, the first processing step is the same as for grey-scale images: the image can be smoothed, sharpened and trimmed, and the contrast and brightness adjusted.

Pixels in multichannel images must be classified, but clearly the process is considerably more complex than for grey-scale images. There are many possible ways to treat such images and only a simplified approach will be discussed here. More details are available in texts such as 'The Image Processing Handbook' (Russ, 1999) or 'Image Analysis in Earth Sciences' (Petruk, 1989).

For many optical images the largest variation between different phases is the brightness of the image, rather than the colour. This is expressed as a significant correlation between the signal in the red, green and blue channels of an RGB image (Launeau *et al.*, 1994, Thompson *et al.*, 2001). That is, if a pixel has an intense red component it is likely that it will also have intense green and blue components. Such images may be usefully converted to other colour systems using GPIP: a useful system is hue, saturation and intensity or brightness (HSI – HSB). Hue is the basic colour, defined by the dominant wavelength of light; saturation is the degree to which the hue is diluted by white light; and intensity or brightness is the overall brightness of the colour. There are a number of other schemes that may be useful also – trial and error is recommended.

The variation between the pixel values for multichannel images (e.g. colours) can be clarified by principal component analysis (PCA; Launeau *et al.*, 1994). The objective of PCA is to decorrelate and concentrate the information carried by many channels. It is equivalent to rotating the axes of the multichannel data in multidimensional space so that the greatest variation lies along the first axis (first component) and the next orthogonal axis (second component) contains the greatest variation in the remaining data. What variation is left lies in the other components. The main disadvantage of this method is that each sample will give axes with different orientations. One solution is to calculate a new set of axes from one sample and apply to all other samples in the same dataset.

The processed multichannel image can be separated into component channels and each thresholded like a grey-scale image to yield binary data using GPIP. It is also possible to select manually the range of colours associated with each mineral. This is done by selecting the colour values associated with the mineral using GPIP (in some programs it is necessary to make a colour 'mask' and to save that as a separate image). Another strategy is to substitute another, stronger colour (e.g. bright green) for the mineral in question and then separate out that new colour. However, much trial and error is needed to ensure a clean separation.

Multichannel data, either RGB, HSI or PCA-transformed components, can be classified in a more elaborate way than just thresholding one channel. Several approaches are possible but all start with training – that is measuring the value of pixels that lie within a known mineral. In the parallelepiped method the upper and lower thresholds are established for each channel, giving a parallelepiped volume for each mineral phase (Figure 2.6a; Launeau *et al.*, 1994). Pixels that fall outside these boxes or in the intersection of two boxes are unclassified. A better method is to define spheres rather than boxes (Figure 2.6b). Where two spheres intersect the contact is planar, like a soap bubble, which eliminates unclassified pixels that lie in two boxes. Pixels outside

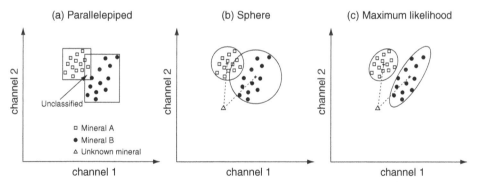

Figure 2.6   Simplified classification of pixels for two channels of multichannel data. (a) Parallelepiped method. Pixels within the parallelepipeds are classified, but those in the area of overlap are unclassified. All other pixels are unclassified. (b) Sphere method. Pixels within spheres are classified. Outside the sphere may remain unclassified or may be classified according to the distance to the nearest sphere centroid. (c) Maximum likelihood method. All points are classified according to their distance to the nearest ellipsoid centroid, normalised by the ellipsoid radius in that direction.

the spheres may be left unclassified or assigned to the sphere with the nearest centroid. Marschallinger (1997) found that the best method was the maximum likelihood classifier (discriminate-function analysis; Figure 2.6c). Results from the training are used to construct ellipsoids instead of spheres. The shape and orientation of the ellipsoid approximates that of the original 'cloud' of pixels. A pixel is classified according to the distance to the centres of the ellipsoids divided by the radius of the ellipsoid in that direction. This method is available in dedicated image analysis programs (see Appendix).

Once the grains have been separated parameters can be evaluated for each entire grain and used to classify the grains into different mineral phases. In addition to the parameters mentioned above, it is also possible to quantify the spatial variations in pixel values within each grain. This is referred to as 'texture', although this term clearly has a wider meaning in this text. So far only the intensity component of the image has been used to evaluate four standard textural parameters called contrast, entropy, energy and homogeneity (Thompson *et al.*, 2001). These all relate to the way in which the intensity varies between adjacent pixels. All these parameters are evaluated for each grain and the results classified using neural networks (Thompson *et al.*, 2001) and genetic programming (Ross *et al.*, 2001). Both give similar results, but neural networks appear to be easier to use and more flexible.

2.6.3.3 Final processing of classified images

The resulting classified images can be manipulated easily in GPIP to remove individual pixels or small groups of pixels, referred to as 'noise'. Touching grains in the classified mineral image can be separated manually by using GPIP also: a line is drawn to separate the grains. If there are many overlapping grains then an automatic procedure may be preferable. These methods usually require that the crystal shape is known. The contours of the grains may be eroded by removing the exterior pixels until the grains are separated. The lost pixels are then added by dilating the grain back to its original size (Russ, 1999). Another choice is 'watershed' algorithm (see Section 2.3). Finally, the sharp re-entrants between touching grains can be detected and the outline segmented at this point (van den Berg *et al.*, 2002). Success rates vary from 50% to 60%, so this method should not be used for samples with large numbers of overlapping grains.

2.6.4 Computer-integrated polarisation microscopy

All automatic processing of transmitted light images currently uses only colours for classification of particles. However, an experienced observer uses other parameters: for example the colour changes produced during rotation of the stage under cross polarisers. These features may be examined using computer-integrated polarisation microscopy (Heilbronner & Pauli, 1993, Fueten, 1997). In this system of microscopy the polariser and analyser rotate together, but the section, and hence the image stays still. Up to 200 images are recorded for different orientations of cross-polarised light. These actual images of the section can be combined and processed to give a series of synthetic images, such as the maximum birefringence (Max) and angle of extinction (Phi). Both Max and Phi give information about the lattice preferred orientation of crystals (Heilbronner & Pauli, 1993, Fueten & Goodchild, 2001). They can also be used to segment the section into separate crystals (Goodchild & Fueten, 1998, Zhou *et al.*, 2004a, Zhou *et al.*, 2004b). An image similar to the maximum birefringence image can be obtained more simply using the Benford plate technique (see Section 2.5.3.1).

2.6.5 *Classified image measurement*

Grain outlines in classified images produced by image analysis may be measured to determine their geometric characteristics. Parameters commonly measured include position of centroid (centre of outline), length and width or major and minor axes of a best-fit ellipse, perimeter, area and orientation of long axis. Many programs can be used for this but the most popular is NIHImage/ImageJ (see Appendix).

2.7 Calculation of three-dimensional data from two-dimensional observations

2.7.1 *Stereology*

The problem of conversion of two-dimensional section data to three-dimensional textural parameters is not simple for objects more geometrically complex than a sphere and there is commonly no unique solution for real data (see below). This subject is treated in a branch of mathematics called stereology. There are a number of reviews and books on the subject, mostly aimed at the biological sciences (e.g., Underwood, 1970, Royet, 1991, Howard & Reed, 1998). Particular aspects of stereology will be discussed in the relevant chapters.

2.7.2 *Stereologically exact 'global parameters'*

There are a number of very useful relationships that can be used to calculate various 3-D 'global parameters' from intersection measurements without any knowledge of the distribution of the phase. That is to say these relationships are correct for all grain sizes, shapes and orientations, provided representative sampling is done (Underwood, 1970, Exner, 2004). It is not even necessary to separate individual crystals or grains. These values are termed stereologically *exact, unbiased* or *accessible* parameters. They are the equivalent of mean values, in that they only describe a limited aspect of the texture. Other textural parameters are termed *biased* or *inaccessible*, but they are still measurable in some materials, if assumptions are made about the crystal shape and fabric, and will be described in later chapters (Underwood, 1970, Howard & Reed, 1998, Brandon & Kaplan, 1999).

The most widely used global parameter is the volumetric fraction of the phase, V_V. The subscript indicates that the volume of phase, V, is normalised to unit volume, V. Delesse (1847) showed that the total area of a phase, A, in a representative section normalised to unit area, A_A is equal to V_V. It may seem counter-intuitive, but the fabric of the rock and the orientation of the section are not important – any plane will give the same area fraction, which is equal to V_V. Crystal intersection areas are commonly measured at the same time as the intersection length and width of the crystals and they can be used to calculate the volumetric phase abundance. In addition, the fraction of a random line that intersects the phase, L_L is also equal to V_V. However, many lines must be measured to give adequate precision. Finally, the fraction of random or equally spaced points that lie on the phase, P_P is also equal to V_V. This is the basis of point counting. Hence

$$V_V = A_A = L_L = P_P$$

Another common global parameter is the interface density, which is the total surface area of grains, S, in a unit volume, S_V. A test line or circle is drawn on the section and the number of intersections with the surface, P, counted (Figure 2.7). This value is divided by the test line length to give P_L.

Alternatively the number of grains intercepted, N, may be counted. This is also divided by the length of the test line to give N_L. Then

$$S_V = 2P_L = 4N_L$$

If there are several phases in a material, then the 'contiguity' of phase α, $C^{\alpha\alpha}$ is the proportion of the interface area shared between like grains. It is determined from the number of test points with $\alpha\alpha$ boundaries, $P_{L(\alpha\alpha)}$

$$C^{\alpha\alpha} = P_{L(\alpha\alpha)}/P_{L(Total)}$$

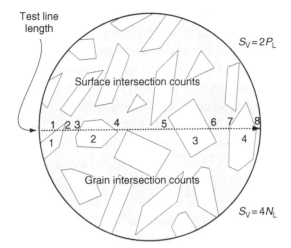

Figure 2.7   Determination of the total surface area per unit volume, S_V from the number of surfaces or crystals intersected. Here, eight surfaces are intersected which gives $S_V = 2*8/\text{length}$ or four crystals are intersected which gives $S_V = 4*4/\text{length}$.

Similarly the 'neighbourhood' is defined as the proportion of interface area shared by two phases α and β, $C^{\alpha\beta}$. It is determined from the number of test points with $\alpha\beta$ boundaries, $P_{L(\alpha\beta)}$

$$C^{\alpha\beta} = P_{L(\alpha\beta)}/P_{L(Total)}$$

Other parameters are the mean linear size of grains, $\bar{L}$

$$\bar{L} = V_V/P_L$$

V_V may be determined from area-based measurements or point counting. It is also possible to determine the mean linear distance between grain centres, $\bar{D}$

$$\bar{D} = (1 - V_V)/P_L$$

The degree of orientation of grains can be determined from the ratio of the number of intersections in orthogonal directions. This is the intercept method, which is discussed in Section 5.4.3.2.

2.8 Verification of theoretical parameter distributions

In natural systems, such as rocks, textural parameters never follow exactly a prescribed distribution law. However, petrologists commonly want to know if a distribution may be approximated by a distribution law, so that the parameters of that law can be used as a substitute for the true distribution. Most of the diagrams in this book are designed specifically for various theoretical distributions. Graphical methods are commonly favoured by geologists to show how well a real distribution corresponds to a theoretical distribution. However, there are a number of mathematical parameters that can also be used to test a hypothesis (Swan & Sandilands, 1995).

A hypothesis is tested using a 'null hypothesis': this is the idea that the data do not differ significantly from that produced by random processes (e.g. Swan & Sandilands, 1995). This hypothesis must be tested using a test statistic. There is a level of significance associated with this hypothesis, commonly 5% or 10%. This means that there is only a 5% chance that the distribution seen occurred by random processes and hence has no significance.

Probabilities can be interpreted as a measure of the times an outcome will occur for a large number of repeated experiments. However, a more useful approach considers probabilities as just an expression of the confidence that a model or theory is correct. When a result is said to be significant or not significant the latter approach is being used.

2.8.1 Chi-squared test

The most widely applied test is the chi-squared (χ^2) test. This is applied to discrete data points.

$$\chi^2 = \sum \frac{(O_j - E_j)^2}{E_j}$$

where j is the class interval number, O_j is the observed frequency of that class and E_j is the expected frequency of that class. Each class should have at least five observations. The significant values of this parameter depend on the number of observations and the required significance level. Tables are available in standard statistical texts (e.g. Swan & Sandilands, 1995).

The chi-squared test is very susceptible to the number of classes (Swan & Sandilands, 1995): The null hypothesis may be more commonly accepted for larger numbers of classes. For a small number of classes failure to reject the null hypothesis (i.e. the data conform to the model distribution) may be due to broad and unimportant similarities between the data and the model (Swan & Sandilands, 1995).

Many other statistical tests have been proposed and the reader should consult more specialised texts (e.g. Swan & Sandilands, 1995).

2.8.2 Straight line distributions

A number of graphs are designed to give a straight line if the textural parameter follows the theoretical distribution. In this case the data can be regressed and the intercept and slope determined. The correlation coefficient r^2 is commonly used to quantify how closely the observed data correspond to a straight line. A value of one indicates a perfect straight line. However, the r^2 parameter alone is not very useful, as the significance of values will depend on the number of points on the graph. Tables are available to assess the level of significance in terms of the number of samples and the value of r^2. A greater problem is that the significance of r^2 will depend on the uncertainty in each point.

A much better assessment of the level of significance can be made if the uncertainty associated with each point is known (Bevington & Robinson, 2003, Higgins, 2006). Ideally the contribution of each point should be weighted for its uncertainty. A goodness-of-fit parameter, Q, can then be calculated that determines the probability that the observed discrepancies between the observed and expected values are due to chance fluctuations (Figure 2.8). A very low value indicates that the discrepancies are unlikely to be due to chance fluctuations. This means that either the model is wrong or the uncertainties have been underestimated or the uncertainties are not normally distributed. If Q is greater than 0.1 then the null hypothesis is rejected and the parameter may

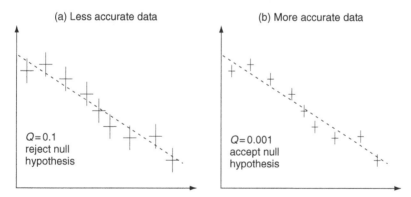

Figure 2.8   Effect of uncertainty on the significance of a distribution. Both diagrams have the same distribution and hence the same value of r^2. (a) Measurements are not very precise and a straight line can be drawn through the uncertainty bars of the points. The value of Q is high enough that the null hypothesis can be rejected. (b) The same points have been remeasured more accurately. A line cannot be drawn through the uncertainty bars and Q is low. Hence, the null hypothesis is accepted and the points do not follow the theoretical distribution.

follow the model distribution. A value of $Q > 0.001$ may be acceptable, because truly wrong models generally have very small values of Q. Values higher than 0.1 do not indicate that the data have more validity. A value close to one may indicate that the uncertainties have been overestimated.

2.9 Summary

Many different analytical methods can be used in textural quantification and it is not always easy to choose the most appropriate method. There must always be a balance between the analytical effort and the expected results: A lengthy method may mean that few samples can be analysed. In some cases the project may benefit from simpler measurements on a larger number of samples. A common strategy for the choice of methods is:

1. **Scale**: ideally, a study should start with the identification of the size scale – e.g. 10 cm megacrysts are best examined on an outcrop, whereas microlites may need electronic methods.
2. **Parameters**: the textural parameters that are needed should be specified – e.g. crystal orientations are not preserved when a sample is dissolved.
3. **Minerals**: it is important to decide what minerals will be examined and if individual crystals can be distinguished by the method.
4. **Equipment**: access to equipment needed for the method must be established, both financially and logistically. If access is not possible then a different project or approach must be chosen.

It should be remembered that if a problem is not tightly constrained then many studies can be done with very simple instruments that are available in almost all geology departments: petrographic microscope, camera, ruler, document scanner and computer.

3

Grain and crystal sizes

3.1 Introduction

One of the most commonly quantified aspects of rock texture is the size distribution of crystals, grains and, more rarely, blocks. Although the first published study of crystal size in igneous rocks was in the early days of petrology (Lane, 1898), the subject really only took off in 1988 after the publication of seminal papers by Marsh and Cashman (Cashman & Marsh, 1988, Marsh, 1988b). In metamorphic petrology the study of size distributions started a bit earlier with the study of garnets by Kretz (1966a). The importance of grain size distribution in sedimentology completely outweighs that of other branches of petrology and is covered in more detail in sedimentological texts (e.g. Syvitski, 1991). Engineers have a considerable interest in the mechanical and chemical properties of materials (Brandon & Kaplan, 1999): grain size and composition play an important role in controlling these properties and hence their quantification is very important (Jillavenkateas *et al.*, 2001, Medley, 2002). Also of interest are pore and bubble sizes in rocks and these are discussed separately in Chapter 7.

3.1.1 What are size and size distributions?

The size of a crystal or grain is a measure of the space occupied by the object. It is a three-dimensional property and can be defined in several different ways, according to the application of the parameter: if the shape of a crystal changes during growth or solution then a volumetric measure of size may be useful. Size based on volume is simple because every grain has a unique volume. However, if we are interested in the growth or solution of crystals where the shape is almost invariant then a linear measure of size is usually appropriate. Hence, the change in size of a crystal can be expressed in terms of linear growth, mm per second, for example. In most geological applications a linear measure of size is used and various definitions are discussed in Section 3.1.2. From here on the term size refers to a linear measure of the space occupied by a grain.

Petrologists commonly describe the mean size of crystals or grains in a rock, without clearly defining what they mean. The definition of size is discussed in Section 3.1.2, but here I want to examine the significance of mean sizes. If a rock contains grains smaller than the resolution limit of the analytical method then the value of the mean size will depend on the resolution (Figure 3.1). In this situation the mean can be defined only if we can describe a size function for that unknown interval. In any case although the mean is a useful parameter, it has only one dimension. Another one-dimensional parameter that is some-times measured is the crystal or grain number density. This is defined as the total number of crystals or grains per unit volume, although some authors use this incorrectly for number density by area. Again, there is a problem if a rock contains grains smaller than the resolution limit of the method. Both these parameters are sub-sets of a more information-rich parameter, the crystal size distribution (CSD), grain size distribution or particle size distribution. This has two dimensions; size and crystal or grain number. It is defined between upper and lower size limits. Outside these limits it is indeterminate, but changing the measurement limits will not change the CSD values within the limits (Figure 3.1).

Finally, some workers loosely use the terms unimodal and bimodal size distributions. The number of modes depends clearly on the variable examined and how it is expressed. For instance, the size variable may be expressed in terms of length or volume of grains: this will clearly change the distribution. In many 'bimodal' rocks there is actually only one mode in the CSD. What is frequently meant by this term is that there are two populations of crystals. However, this should be verified by other means, such as chemical or isotopic composition, as single populations may appear visually to be multiple populations.

3.1.2 Linear size definitions

The linear size of a grain has many different definitions. This is compounded by a common confusion between true size in three dimensions and size in intersections or projections of grains. Three-dimensional linear size definitions are listed here in order of typical decreasing numerical value.

1. The greatest distance between any points on the surface of the grain.
2. The length of the smallest parallelepiped that can be fitted around the grain.

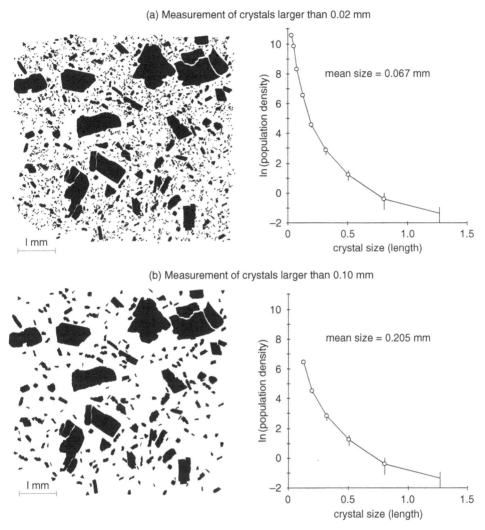

Figure 3.1   Effects of size limits on values of mean crystal size and crystal size distributions: example of an andesite from Soufriere Hills, Montserrat (Higgins & Roberge, 2003). The original data have been slightly modified for this diagram. (a) All crystals larger than 0.02 mm have been counted. Crystals smaller than this have been observed but not quantified. Hence, the CSD is undefined below 0.02 mm. The mean size is 0.067 mm. (b) All crystals larger than 0.10 mm have been measured. The CSD for sizes greater than 0.1 mm is identical to the upper diagram, but the mean crystal size, 0.205 mm, is almost three times greater. Hence, if a rock contains crystals of all sizes then the mean size is meaningless – it just depends on the measurement size limits.

3. The mean projected height (mean calliper diameter) of the grain. This can be visualised at the mean height of the shadow of the grain when it is illuminated from all different angles. This is commonly used in stereology.
4. The length of the major axis of a best-fit ellipsoid. This ellipsoid has the same volume and moment of inertia as the grain.
5. The diameter of a sphere of equivalent volume, or more rarely the edge of a cube of equal volume.

Other size definitions used for unconsolidated sediments are described in Section 3.3.1. For grains close in shape to a sphere all definitions give similar sizes. The difference between the measures of size increases with the anisotropy of the grain. Definitions 2, 3 or 4 are generally similar and should be used. The definition of size used should always be specified in publications.

3.2 Brief review of theory

A wide range of processes and constraints can affect the crystal or grain size distribution of rocks, hence the interest in this parameter. For igneous rocks we must consider the kinetic process of nucleation and initial growth (Figure 3.2). The driving force for kinetic processes can be expressed as the undercooling or supersaturation of the system (see Section 3.2.1.2). The resultant crystal populations may be modified by mechanical processes such as sorting and mixing. Once the kinetic driving force is reduced, the initial or modified crystal populations will start to equilibrate, although complete equilibrium is never achieved. This process involves the minimisation of the total energy of the crystal population. The importance of distinguishing kinetic and

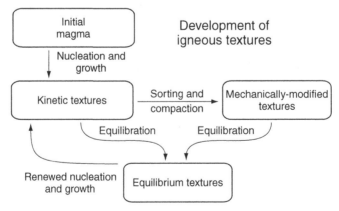

Figure 3.2   Development of igneous textures – role of kinetic, mechanical and equilibrium processes.

equilibrium effects cannot be overstated – it is frequently a source of significant confusion in the interpretation of CSDs. Melting of igneous rocks will probably be more controlled by equilibrium textural processes, as it is easier to destroy a crystal lattice than to make one. But again we must consider mechanical effects on the crystal population. Kinetics and equilibrium are also important in metamorphic rocks, but deformation clearly plays a more important role than in igneous rocks. Finally, I will consider the most generally applicable constraint on grain and crystal populations, closure: this is the fact that a rock cannot contain more than 100% crystals or grains. As in geochemistry, this constraint is commonly acknowledged but it is ignored (Higgins, 2002a).

3.2.1 Primary kinetic crystallisation processes

In many branches of geology the roles of kinetics and equilibrium are important, and that is certainly the case here: kinetics will control the speed at which crystals can nucleate and grow (or dissolve) and equilibrium will dictate the final amount of the phases. The most common way of expressing the kinetic effect is in terms of undercooling: a certain overstepping of equilibrium is necessary to force the nucleation and/or growth of a new phase. However, the amount of undercooling encountered during the crystallisation of rocks is probably strongly dependent on the crystallisation environment and minerals present, and is not well established so far.

The fundamental functions that control growth textures are simply expressed: during solidification the nucleation rate is a function of time $J(t)$, whereas the growth rate is a function of both time and crystal size $G(t,l)$. If both functions are known, then the evolution of the crystal size distribution with time, $n(t,l)$ can be determined. Generally speaking we are more interested in the inverse problem: that is we have measured the final crystal size distribution and we want to know the nucleation and growth rate variations during solidification. There is no unique solution to this problem, but many attempts have been made by making assumptions about the form of $G(t,l)$ and $J(t)$ (e.g. Brandeis & Jaupart, 1987, Marsh, 1988b, Lasaga, 1998, Marsh, 1998, Zieg & Marsh, 2002).

It is commonly overlooked that significant undercooling is not necessary for crystallisation: equilibrium requires that the total volume of crystals will increase during cooling. The texture of the crystals responds to equilibrium effects to minimise the total energy of the crystal population. This process is common and has many names: the effect on crystal size is referred to here as coarsening (see Section 3.2.4). Hence, the final texture of most igneous rocks will reflect some combination of kinetic and equilibrium effects.

3.2.1.1 Nucleation of crystals

In a liquid, such as magma or an aqueous solution, atoms are continually being added or removed from clusters that have a lattice arrangement similar to that of major mineral phases. Addition of new atoms creates bonds that reduce the energy of the cluster. However, bonds at the surface of the cluster are unsatisfied and there is a net increase in energy. The sum of these two components gives the total energy of the cluster. With increasing cluster size the total energy rises to a peak and then descends (Figure 3.3a; e.g. Dowty, 1980, Markov, 1995). When one of these clusters exceeds a critical radius, r^*, then it is energetically stable and a crystal nucleus is formed. Addition of further atoms to the nucleus is energetically favoured as it reduces the total energy. This process is termed homogeneous nucleation. The driving force for nucleation is related to the undercooling of the system (see next section). At low undercooling the energy barrier is not breached, no clusters reach the critical size and the nucleation rate is zero (Figure 3.3b). With increasing undercooling, the nucleation rate increases rapidly to a peak and then diminishes.

Heterogeneous nucleation occurs on the surface of existing host grains or bubbles, or on defect structures within crystals. The surface energy of the cluster is lower, hence both the critical radius and energy barrier are lower (Figure 3.3a). Hence, formation of stable nuclei occurs at lower undercooling than for homogeneous nucleation. Mismatches between the crystal lattices of the new and existing grains will produce elastic stresses that store energy. Hence, the actual undercooling will generally depend on the degree of crystallographic similarity between the host and the new phase. This is expressed as the dihedral angle, θ (see Section 4.2.3; Figure 3.3a).

The necessary undercooling is obviously smallest when the new and host crystals are mineralogically identical and in the same orientation. However, in this case the greater ease of nucleation will be balanced by the lower concentration of the mineral component in the magma surrounding the growing host crystal.

Crystal nuclei are too small and unstable to be observed – we can record them only if there has been an interval of growth. Hence, it has been difficult to study nucleation. Many authors assume that magmas are never completely liquid and that there is always the possibility of heterogeneous nucleation. However, as was shown above, the energy difference as expressed by the dihedral angle may be sufficiently high that homogenous nucleation occurs even in the presence of solid phases.

New crystals may also form within existing crystals, for instance during exsolution related to cooling of feldspars. Such crystals can form

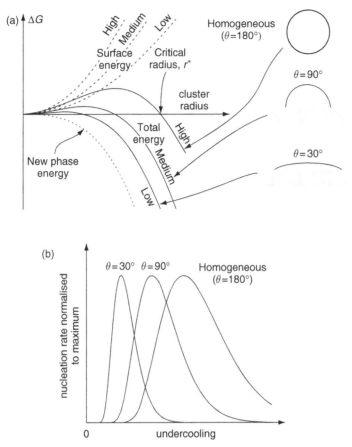

Figure 3.3 (a) Total energy of an atomic cluster against cluster radius. The total energy of an atomic cluster is the sum of the surface energy and the new phase (lattice) energy. For homogeneous nucleation the surface energy is high and the critical radius is large. For heterogeneous nucleation the surface energy is lower giving a lower critical radius. Two cases are shown here which are expressed by dihedral angles of 90° and 30°. Although homogeneous and heterogeneous nucleation processes are commonly presented as opposites, it is clear that there is a continuum between these two situations. (b) For homogeneous nucleation at small degrees of undercooling the normalised nucleation rate is zero. As undercooling increases, it increases rapidly to a peak and then diminishes. In geological situations most nucleation probably occurs to the left of the peak. Heterogeneous nucleation reduces the undercooling necessary for nucleation: a lower energy barrier (ΔG) leads to lower dihedral angles and lower undercooling.

discontinuously by nucleation and growth, in the same way as for nucleation in fluids or on surfaces. However, heterogeneities in crystals may also arise continuously by a process called spinoidal decomposition (McConnell, 1975). Compositional waves form spontaneously in the mineral lattice: the wavelengths are short at low temperatures, but become longer with increasing temperatures. Such zones may later transform into separate phases which can be observed as regular exsolution patterns of contrasting phases. The whole process is not dissimilar to geochemical self-organisation (Ortoleva, 1994).

3.2.1.2 Growth and solution

At its simplest, crystals grow if this results in a reduction in the total energy of the crystallising system, that is the population of crystals and the material from which they are crystallising. Similarly, crystals will dissolve also if this results in a reduction in the total energy. We are principally interested in crystal growth rates: that is the change in size of the crystal with time. Unfortunately, although time is the essence of geology, the timing of crystal growth is generally difficult to measure in the earth. Therefore, experiments are an important complement to natural studies for determining growth rates. Crystal growth rates are very important in many industrial processes and hence there is an immense amount of work in this subject. The problem for geologists is to determine what aspect of this work is applicable to our field. Some have seen the range in possibilities, found them wanting and created their own arbitrary growth laws (e.g. Eberl *et al.*, 1998, Eberl *et al.*, 2002). Here, I will present a very brief overview of what I consider to be important. More complete treatment of the subject is available in a number of books and articles (e.g. Markov, 1995, Marsh, 1998, Jamtveit & Meakin, 1999).

Crystal growth rates depend on the magnitude of the driving force, but the exact relationship is controlled by the growth mechanism. The driving force for crystal growth is equal to $\Delta\mu/kT$, where $\Delta\mu$ is the chemical potential difference between the crystallising phase and the host fluid (magma, liquid or gas), k is the Boltzmann constant and T is the temperature. $\Delta\mu/kT$ may be expressed in many different ways, but the most useful for us is for solutions in terms of concentrations or enthalpy and temperature (e.g. Lasaga, 1998):

$$\frac{\Delta\mu}{kT} = \ln\left(\frac{C}{C_0}\right) = \frac{\Delta H_{sol}(T_{sat} - T)}{kTT_{sat}}$$

where C and C_0 are the real and equilibrium concentrations of the solute; ΔH_{sol} is the positive enthalpy of solution; T_{sat} is the saturation temperature (melting point; mineral liquidus temperature). It is assumed that the system is ideal and that the enthalpy of the transition is independent of temperature. In most applications it is easiest to use the undercooling as a measure of driving force. However, if we want to look at the relationship between crystal size and growth rate then it is more convenient to use concentration terms.

The mechanism of crystal growth is by addition of atoms to the surface. The rate of addition can be controlled in two ways (e.g. Baronnet, 1984, Lasaga, 1998);

- By the processes of addition of atoms to the surface: interface-controlled growth.
- By the transport of the crystal components towards the surface and unwanted material and latent heat away from the surface: transport-controlled growth. The actual movement is controlled by diffusion and movement of the fluid with respect to the crystal (advection).

The growth shape of a phase depends partly on its physical isotropy or anisotropy: a physically isotropic material has the same distribution of atoms in every direction. Isometric or cubic minerals are optically isotropic, but not physically isotropic. If crystals were physically isotropic then they would grow in ideal circumstances as spheres: liquids are isotropic and do grow in this way. However, all crystals are physically anisotropic: hence, the nature of the crystal–host interface is different in every direction. Of course, the amount of anisotropy in a crystal is variable and is expressed relative to the contrast between the crystal and its host. This contrast varies with temperature: below the 'roughening transition' temperature crystal faces grow by spiral growth or two-dimensional nucleation (Figure 3.4). They keep their orientation and continue to grow as macroscopically flat faces. Above the roughening transition the energy difference between the crystal and the host dominates: the crystals grow with rounded surfaces, without facets. In the

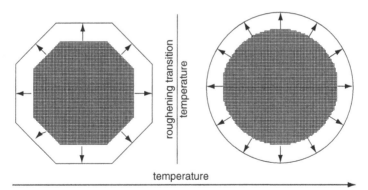

Figure 3.4 Roughening transition (e.g. Baronnet, 1984, Lasaga, 1998, Bennema *et al.*, 1999). Below this temperature crystals grow with facets; above it they grow with rounded surfaces. This transition resembles a phase change. It should be remembered that the roughening transition temperature may be higher than the saturation temperature (liquidus) of a mineral phase, in which case the mineral will always crystallise with faces where it does not impinge on other surfaces.

geological environment most crystals grow with facets, although these may be lost where adjacent crystals impinge. Curved faces do occur (e.g. Heyraud & Metois, 1987), but those produced by solution can generally be distinguished from those produced during growth by examination of internal growth zones.

Supply of crystal materials depends on several factors: the chemical contrast between the crystal and its environment and rate of transport of atoms. Transport includes both diffusion and the relative physical movement of crystal with respect to the host. The first term is easier to treat: if a crystal has exactly the same composition as the 'mother' material then no diffusive transport is necessary. An example would be the crystallisation of a pure substance, such as water. Pure substances are rare in geology: we are normally concerned with solutions such as magmas.[1] Another example is the enlargement of a grain by grain-boundary migration: here, the original and final materials have the same composition and the transport distance is short.

In more complex situations transport of atoms is controlled by numerous factors:

- Physical state of the host material: solid or liquid. Diffusion is much faster in a fluid phase[2] than a solid phase.
- Composition of the host material. The overall composition is expressed as the viscosity. Atoms move more slowly in viscous materials. However, all atoms do not move at the same rate: generally, small ions move faster than large atoms. Addition of material to chemically complex minerals requires diffusion of many different ions, and the overall process is controlled by the slowest atomic species. Addition of water to magma will reduce the viscosity and increase the diffusion rates of all species.
- The temperature of the system is extremely important: higher temperatures increase diffusion rates. However, pressure has little effect as solids are almost incompressible.
- Relative movement of fluid (advection) from which the crystal is growing with respect to the crystal. This replenishes the boundary layer that was depleted in the crystal components by crystallisation and enriched in the rejected components.

3.2.2 Kinetic textures

Crystal populations produced by kinetic processes alone are most commonly modelled using the Avrami equations (e.g. Lasaga, 1998). However, the solution of this equation requires knowledge of the variation of the nucleation rate with time, $J(t)$, the growth rate with time and crystal size, $G(t,l)$. Both of these are generally unknown and various models of their behaviour have been proposed. The most common assumption is that the growth rate is independent of size and hence this variable can be simplified, although Eberl *et al.* (1998) have proposed that it is proportional to size. It cannot be overstated that independence of growth rate and size is for kinetic growth only: it is well established that during coarsening (equilibration) the growth rate is size dependant (e.g. Voorhees, 1992). Both Gaussian (Lasaga, 1998) and exponential (Marsh, 1998) models for the variation of growth and nucleation rates with time have been proposed. One of the most popular population models uses a steady-state approach and will be discussed first.

3.2.2.1 Steady-state crystallisation models

The nucleation and growth of crystals is a very important industrial process and much work has been done in this field. One important type of industrial reaction vessel has a continuous input and output of material, with crystallisation in the vessel. The input material may contain crystals. Crystallisation occurs in the reactor; hence the output material is richer in crystals than the input material. The classic study of CSDs in this system by Randolph and Larson (1971) was adapted to magmatic systems by Marsh (1988b; Figure 3.5). In both industrial systems and topologically similar open magmatic systems there is a linear relationship between the natural logarithm of the population density at size L and that size. For one phase in the system:

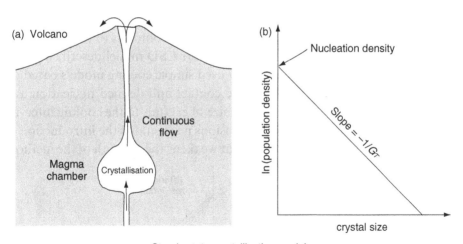

Steady-state crystallisation model

Figure 3.5 (a) Marsh (1988b) proposed that a continuously erupting volcano approximates a steady-state reactor. Magma with or without crystals enters the magma chamber from below and partial crystallisation occurs in the chamber. An equal quantity of magma is withdrawn and erupts. (b) The CSD of a phase in the magma chamber is a straight line on graph of ln (population density) against size.

$$n'_V(L) = n'_V(0)e^{-L/G\tau}$$

where $n'_V(L)$ is the population density of crystals for size L, $n'_V(0)$ is the final nucleation density, G is the growth rate and τ is the residence time. The characteristic length, C, is defined by:

$$C = G\tau$$

It is equal to the mean length of all the crystals in a straight CSD that extends from zero to infinite size. This distribution is linear on a graph of ln(population density) versus size (L). The intercept is $\ln(n'_V(0))$ and the slope is $-1/C$. For steady-state systems the characteristic length equals the residence time multiplied by the growth rate. There are few geological systems that exactly resemble the original industrial system, but the approach of Marsh (1988b) gave considerable impetus to the study of CSDs.

This crystallisation model predicts the CSD of a system in a steady state, which is a single CSD. In geology we are more interested in the dispersion of parameters in systems. What sort of variation can we then expect in such a system? If the residence time is constant, but the nucleation density is increasing then the CSDs will be parallel (Figure 3.6). Increasing nucleation density could be produced by increasing undercooling. Increasing residence times for constant nucleation density will make the CSDs swing around a point on the vertical axis. Such a situation could occur if the magma chamber is expanding. Any combination of these two models is also possible. There is, however, no observational evidence so far that any geological system actually conforms to these models.

Zieg and Marsh (2002) have taken the linear CSD model described above and linked it to cooling equations. They used simple cooling models based on one-dimensional cooling across a single contact and defined nucleation and growth parameters that relate the mean size of crystals to the cooling interval. Application of this theory to actual intrusions necessitated the introduction of an arbitrary growth law, which somewhat weakens the strength of the method.

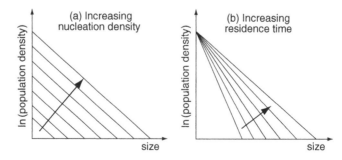

Figure 3.6 Theoretical dispersion of CSDs in a continuously fed crystallisation system with (a) variable nucleation density and (b) residence time.

Changes in the crystallisation parameters during solidification, such as cooling rate, may change the characteristics of the CSDs. In general the system will take time to accommodate the new conditions and during that time the CSD will be curved on a classic CSD diagram (Marsh, 1988b, Maaloe *et al.*, 1989, Armienti *et al.*, 1994). Such a transition will change the nucleation rate of the magma and initially produce a steeper part of the CSD at small sizes. However, with time the increased nucleation rate will propagate through the CSD as a step, until finally the CSD is again flat from end to end. Cashman (1993) proposed that primary curved CSDs could be produced at a constant cooling rate if there is a change in the nature of the phases precipitated. In this case there must be crystallisation over a significant period of time in the two-phase field.

3.2.2.2 Batch crystallisation: Marsh (1998) model

Obviously the steady-state approach of Randolf and Larson (1971) cannot be applied directly to closed-system crystallisation. However, Marsh (1998) also examined closed systems and discovered that a similar correlation can be produced under certain circumstances, although for high crystal contents the CSD is not linear for small crystals (Figure 3.7). In such a system the logarithmic–linear correlation is produced by exponentially increasing nucleation density with time, as would be produced if the undercooling of the magma

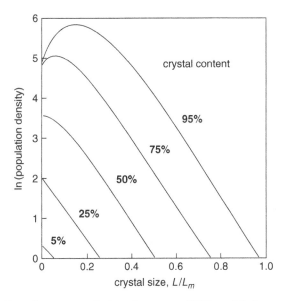

Figure 3.7 Development of closed-system CSDs with increasing crystal content. Crystal sizes have been normalised to that of the largest crystal. The turn-down for small sizes reflects the lower magma volume and hence nucleation rate at high crystal contents. After Marsh (1998).

was increasing linearly. Just as for the case of the open system, a length can be derived from the slope of the CSD. However, this 'characteristic length' does not have a clear physical meaning, as was the case for the open system. Experimental or natural validation of these models is lacking – the only system studied in detail that might replicate closed-system CSDs, the Makaopuhi lava lake, had CSDs with constant intercept and variable slopes (see Section 3.4.1.1; Cashman & Marsh, 1988).

3.2.2.3 Other kinetic population models

Lasaga (1998) has explored the CSDs that are produced by Gaussian-shaped variations of nucleation and growth rates with time. He found that the growth rate must peak before the nucleation rate reaches its maximum otherwise totally unnatural CSDs are produced. He also showed that few CSD models produce straight CSDs on a classical CSD diagram and that minor variations in the parameters of the rate variations can produce significant changes in the CSDs.

Eberl, Kile and co-workers have proposed that kinetic growth rate is proportional to crystal size (Eberl *et al.*, 1998, Kile *et al.*, 2000, Eberl *et al.*, 2002). They based this idea on their observation that many CSDs are lognormal. However, in many of their studies it is not clear if the CSDs are truly lognormal when the data have been stereologically corrected and lower analytical cut-off limits are respected. Approximately lognormal CSDs do exist (Bindeman, 2003) and are easily produced during equilibration (see Section 3.2.4) and this is most likely the case for their observations.

Maaloe (1989) has proposed that many CSDs are exponential and that a growth probability model can explain this distribution. However, this effect was not enough to explain the curvature of his CSDs and hence he also invoked a two-stage model of crystallisation.

3.2.3 Mechanically modified textures

Populations of crystals can be modified by mechanical processes, such as compaction, sorting and mixing. Some of these processes are closed, in that the total mass of the system is unchanged, others are open to the addition of crystals and magma.

3.2.3.1 Compaction and filter pressing

A common process that can change CSDs is compaction (e.g. McBirney & Hunter, 1995, Hunter, 1996, Jerram *et al.*, 1996, Higgins, 1998). This process starts as mechanical compaction, in which reorganisation of the grains enables

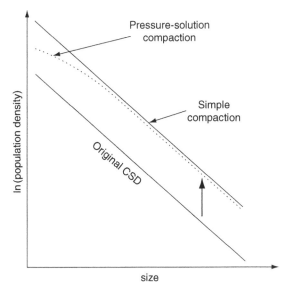

Figure 3.8 Simple compaction changes all the population density values evenly. Pressure-solution compaction will probably result in the preferential resorption of small crystals and will hence have similarities with coarsening processes (Higgins, 1998).

pore fluid to be squeezed out. This will change the population density of each point on the CSD, pushing it upwards evenly (Figure 3.8). It will result in an increase in the intercept, but not the slope of the CSD. For instance, 50% compaction will increase the intercept by 0.7 (= ln(1/0.5)). All minerals that were present during compaction should show the same displacement of their CSDs. It is possible that the finest grains may be removed with the fluid – this will produce a lower population density at the left end of the CSD. Mechanical compaction is limited to a maximum of about 50% by the original coherence of the magma and does not lead to the development of a significant foliation (Higgins, 1991). Filter pressing is the same process except that we are more concerned with the expulsed liquid.

Higher degrees of compaction must involve changes in the shape of the grains. Hunter (1996) has reviewed the different mechanisms by which crystals can change shape during compaction. For cumulate rocks with a small amount of melt, melt-enhanced diffusive creep will be the most important mechanism. Material will be dissolved from surfaces with higher energies, such as those in contact with other grains. This is commonly termed the pressure-solution effect (Figure 3.9). Meurer and Boudreau (1998) proposed a compaction model where mechanical rotation of grains is accompanied by pressure solution of the ends of grains not orthogonal to the direction of maximum

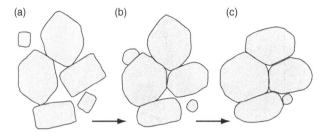

Figure 3.9   Textural changes during pressure-solution (textural equilibrium) compaction. After Hunter (1996). (a) Initial impingement texture. (b) Early pressure solution. (c) Porosity eliminated by pressure solution.

stress. This process could lead to much larger degrees of compaction and the development of a significant foliation. Although this model has not been quantified, it has many aspects in common with coarsening – parts of grains go into solution because they have higher surface energy. Higgins (1998) proposed that the CSDs produced by pressure-solution compaction are similar to those produced by coarsening (Figure 3.8). However, the two processes are not exactly equal as crystals of all sizes can be dissolved by pressure solution.

Pressure solution is driven by minimisation of the total energy of the system as is the coarsening process. Hence, if CSDs are modified in the same way then there should be a correlation between the characteristic length of the CSD and the quality (degree) of foliation. Higgins (1998) found that a few samples showed this effect, but most did not. Hence the process may only be locally important.

There is another effect by which this process may be recognised: during compaction-pressure-solution the ends of crystals not orthogonal to the direction of compaction will be dissolved and hence the shape anisotropy will no longer correspond to the lattice orientation.

Although compaction pressure solution may be able to produce a significant foliation, flow of crystal-rich magma is a much more efficient process (e.g. Higgins, 1991, Ildefonse *et al.*, 1992, Nicolas, 1992, Nicolas & Ildefonse, 1996). The two processes can be distinguished if the magma contains prismatic crystals as flow produces a lineation. In many cumulate rocks plagioclase is measured and this is generally tabular, hence without a lineation. Simple magmatic flow does not change the CSDs, and hence its action can only be revealed by textural measures of crystal orientation. Nicolas and Ildefonse (1996) have proposed that pressure-solution plays a significant role during flow of magma by removing obstacles and entanglements. In this case it would affect CSDs in the same way as compaction-pressure solution.

3.2.3.2 Crystal accumulation

Crystal accumulation is the process of the separation of crystals from magma. There are many different mechanisms that can achieve this: filter pressing (gas exsolution), flowage differentiation (Bagnold effect), gravity separation and crystallisation on conduit walls. Although he did not model this process, Marsh (1988b) suggested that accumulation of crystals by gravity will produce curved CSDs on a classical CSD diagram. Higgins (2002b) has developed a simple model for crystal accumulation based on Stokes' law (Figure 3.10).

For the purposes of modelling, this process is considered here only in its simplest form – the separation of crystals and liquid in response to differences in density. Stokes' law is used to calculate the CSDs in a zone of accumulation of crystals. Many magmas are not Newtonian, especially at high crystal concentrations, but their yield strength under different conditions is not well known (Naslund & McBirney, 1996). In addition, Stokes' law is for spheres. Hence, Stokes' law is not totally applicable in a magma chamber and this model should be viewed as a point of departure, which is probably more accurate for the initial stages of the process when crystal concentrations are low.

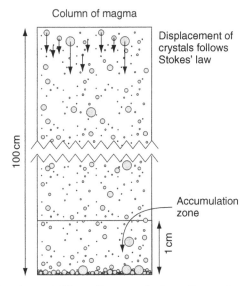

Figure 3.10   Simple modelling of accumulation of crystals using Stokes' law (Higgins, 2002b). The position of each crystal after a discrete time interval is calculated from Stokes' law. The crystals are allowed to accumulate in a zone at the bottom of the column. Flotation of crystals could be modelled in the same way.

A population of crystals is placed in a column of magma and allowed to settle under gravity. Their vertical positions at subsequent times are calculated using Stokes' law:

$$v_t = \frac{2g\Delta\rho r^2}{9\eta}$$

where v_t = terminal velocity, g = gravitational constant, $\Delta\rho$ = density difference between the magma and the crystals, r = crystal radius and η = viscosity. The CSDs of those crystals present at the bottom of the column is that of the cumulate rock. For simplicity, a fixed accumulation zone is defined that can hold all the crystals in the column. A low initial concentration of the crystals ensures that the collection zone never fills up and that the small number of crystals still in the magma does not have a large effect on the cumulate CSD.

This model was applied by Higgins (2002b) to the settling of plagioclase and olivine in a basaltic magma (Figure 3.11). The model begins with a population of spherical olivine and plagioclase crystals that has a straight-line CSD with slopes and intercepts similar to those observed in the Kiglapait intrusion (Higgins, 2002b), but at only 1% of the crystal concentration. These crystals are initially randomly distributed in a one-metre-high column. Initially the CSDs remain linear and rotate about the intercept. After one hour 35% of the olivine but only 4% of the plagioclase has been precipitated. Later, the right ends of the CSDs reach a limit as all the crystals have been precipitated. The linear section then propagates towards the origin. At later stages the CSD has a pronounced turn-down for small crystals, but eventually becomes a straight line parallel to the original CSD. At the end the whole process is equivalent to simple compaction. In a real situation, for example at the base of a magma chamber, crystal settling is never allowed to go to completion: new magma will sweep in displacing the older, fractioned magma. Hence, the early part of this model is probably the most applicable.

3.2.3.3 Mixing of magmas and crystal populations

Mixing of magmas is a very common process – many plutonic and volcanic rocks show abundant evidence for this process. The term mixing is usually reserved for the extreme limit of the process that produces a homogeneous product. Mingling refers to a less extreme process where the two components can still be distinguished. However, this is just a question of scale. CSDs are a bulk measure and hence will sample macroscopic regions that may not be completely homogeneous.

Higgins (1996b) showed that addition of two straight CSDs with contrasting slopes and intercepts will give a CSD with a steep slope at small sizes and a

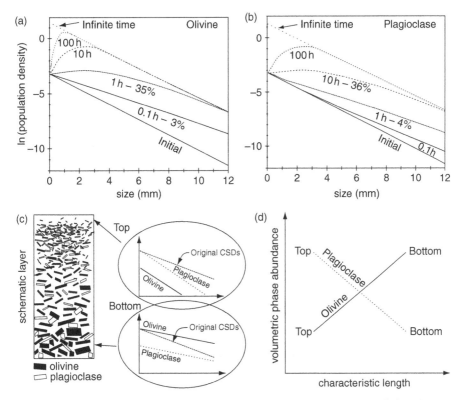

Figure 3.11 Modelling of accumulation of plagioclase and olivine in a basaltic magma using Stokes' law (Higgins, 2002b). The following parameters were used: viscosity $= 400\,\text{Pa\,s}$, magma density $= 2620\,\text{kg\,m}^{-3}$, olivine density $= 3500\,\text{kg\,m}^{-3}$, plagioclase density $= 2700\,\text{kg\,m}^{-3}$. The magma column was one metre high. The CSDs are for the bottom 1 cm. (a, b) The evolution of plagioclase and olivine CSDs from an initial, straight CSD. (c) Schematic effect of settling of plagioclase and olivine in the column. (d) Effect of settling on the amount of a phase (volumetric phase abundance) and the characteristic length ($-1/\text{slope}$) of the CSD.

more gentle slope at larger sizes. Summation of the two linear CSDs gives a curved CSD with two linear segments because of the logarithmic scale of the population density axis (Figure 3.12). It is clear that the slope of the original CSD of both magmas can be recovered from the linear segments even when one component is only 1% of the total. However, it should be noted that although mixing of two populations of crystals will not change the slope of their CSDs, the associated intercepts will be lower than that of the original magmas. Therefore, if the Marsh (1998) model is accepted then residence times can be determined for each population, but no meaning can be attached to the value of the intercept, unless it can be corrected for the effects of dilution by the

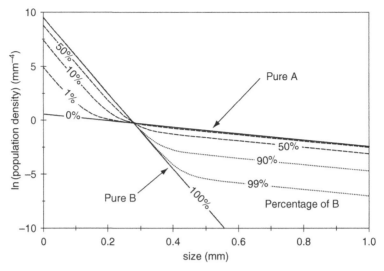

Figure 3.12 Addition of two linear CSDs (after Higgins, 1996b). The original slopes of the two components are preserved for large and small crystals, but not their intercepts.

other magma. Such corrections can be determined from the mixing proportions or the initial crystallinity of the magma. Further growth of crystals following mixing will generate a third population of crystals, represented by another linear segment of the CSD at the left of the diagram. The existing crystals will grow rims and their CSDs will be displaced towards the right of the diagram, but the slopes of the linear segments will be preserved. It should be noted that the deconvolution method described above can also be applied to multiple populations of crystals that grew sequentially from the same magma, if such processes produce straight CSDs.

If two magmas are mixed together then the volumetric proportion of a phase V_{total} in the final product can be determined from (Higgins, 1996b):

$$V_{total} = V_A X_A + V_B X_B$$

where V_A, V_B are the volumetric abundances in end-member magmas A and B and X_A, X_B are the volumetric proportions of the magmas. This equation can be recast as:

$$X_A = \frac{V_{total} - V_B}{V_A - V_B}$$

Therefore, if the initial volumetric phase abundance of both magmas can be estimated and the final volumetric phase abundance of the mixture is known, then the proportions of the two magmas can be calculated. Unfortunately, the initial volumetric phase abundance of both magmas is not determined easily. The apparent volumetric phase abundance of the two magmas can be calculated from the modelled slope and intercept of the CSD for each crystal population and the crystal shape. However, this is not the initial value before mixing, but a lower value produced by 'diluting' the initial volumetric phase abundance with the other magma.

3.2.4 Textural effects of equilibration

At the surface of a crystal there are unsatisfied bonds which have a greater energy than those within the crystal: this excess energy is minimised as the rock texture equilibrates. Although complete equilibrium is never reached the texture will tend towards a population of ever larger, even-sized crystals. Crystal populations produced by kinetic processes, even if modified by mechanical processes, are far from equilibrium as they contain crystals with a wide range of sizes: smaller crystals will have a higher surface area and hence energy per unit volume than larger crystals (Figure 3.13). Equilibration will reduce the total energy, which in a defect-free crystal is related to the total surface area of the crystals. The process generally starts at grain boundaries where the dihedral angles will adjust to uniform values (see Section 4.2.3). The process will continue into the rest of the crystal boundaries, giving them an

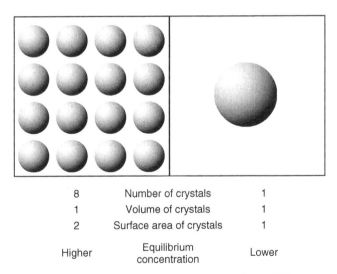

8	Number of crystals	1
1	Volume of crystals	1
2	Surface area of crystals	1
Higher	Equilibrium concentration	Lower

Figure 3.13 Spherical crystals of the same phase of two different sizes. Both populations have the same total volume but different crystal numbers and surface areas. Equilibration will tend to reduce the number of small crystals and assemble their material into larger crystals.

even curvature. Finally, the size population of the crystals will be affected by the solution of small crystals and the growth of larger crystals, which is discussed here. There are aspects of crystals other then size that will also contribute to excess energy: complex crystal boundary shapes and lattice defects are probably the most important. Minimisation of the energy associated with these structures may also power coarsening processes, although initially the process may take a different form (see Section 4.2.2).

The process of textural equilibration is commonly observed and carries many names, some with slightly different definitions: *coarsening* (Higgins, 1998), *textural maturation, textural equilibrium* (Elliott & Cheadle, 1997), *Ostwald ripening* (e.g. Voorhees, 1992) *crystal aging* (Boudreau, 1995), *competitive particle growth* (Ortoleva, 1994), *annealing* and *fines destruction* (Marsh, 1988b). Metamorphic petrologists also use a wide range of terms for this general process: *recrystallisation* (a somewhat ambiguous term), *grain-boundary area reduction* and *grain-boundary migration*. The term coarsening will be used here as it applied to CSDs.

3.2.4.1 Energy considerations

Coarsening is an important process in metallurgy, ceramics and other industrial fields (see the review by Voorhees, 1992). Most authors consider that it is important in metamorphic rocks (Nemchin *et al.*, 2001), although there are some dissenters (e.g. Carlson, 1999). The role of coarsening in the development of igneous rocks is much less well established but seems to be important in plutonic and some volcanic systems (e.g. Hunter, 1996, McBirney & Nicolas, 1997, Higgins, 1998).

The total surface energy of a grain is related to its area by the interfacial free energy, γ. The equilibrium concentration of an essential element for a crystal of radius r, $C_e(r)$, in the material surrounding the crystal will have to reflect these differences in energy: smaller crystals will have higher values than larger crystals. This can be calculated with the Gibbs–Kelvin equation for isotropic materials (e.g. Marqusee & Ross, 1983):

$$C_e(r) = C_e(\infty)e^{\alpha/r}$$

where

$$\alpha = \frac{2\gamma w}{kT}$$

and w is the unit cell volume of the mineral, k is the Boltzmann constant and T is the temperature in K. The value of $C_e(r)$ decreases slightly at higher temperatures. However, the most important control, apart from radius, is

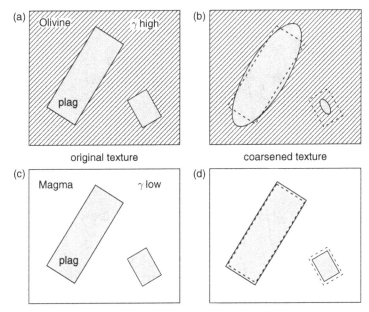

Figure 3.14   The textural expression of differences in interfacial energy, γ, between two phases during equilibration. (a) If plagioclase crystals are enclosed by an olivine oikocryst, then the interfacial energy difference will be large. (b) With equilibration the plagioclase chadocrysts will become rounded and coarsened. (c) If plagioclase crystals are immersed in a basaltic magma then the interfacial energy difference is smaller, as the atomic arrangements are less contrasting. (d) With equilibration the plagioclase crystals will be less coarsened and rounded.

the interfacial energy, γ. This parameter can have a significant effect on rock textures (Figure 3.14).

If crystals of different sizes are adjacent then the equilibrium concentrations of essential elements near the surfaces of the crystals will be different. Clearly this concentration gradient cannot be maintained and diffusion will tend to even out the concentrations. Once the concentration of an element is changed the crystal will either grow or dissolve. Normally, the smaller crystals will dissolve and the large crystal will grow. All the parameters of this equation are well known except the interfacial energy. This depends on the nature of the surrounding material. Relative values can be derived from the dihedral angles of crystals (see Section 4.2.3), but absolute values are much less well known: Nemchin *et al.* (2001) reported a range of a factor of 80 for the interfacial energy of zircon.

For igneous systems there is another way of looking at the problem. Variation of the rates of nucleation and growth with undercooling are well known (Figure 3.15). Coarsening can only occur where the nucleation rate is

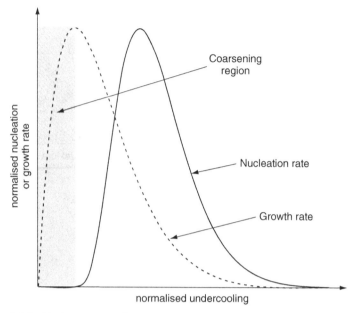

Figure 3.15   In igneous rocks coarsening occurs at low undercooling, close to the liquidus, where the nucleation rate is zero, but the growth rate is significant.

zero, but the growth rate is significant. This is close to the liquidus of that mineral. This can also be visualised in terms of the critical radius of atomic clusters: close to the liquidus even very large, macroscopic clusters are not stable.

Material is transferred from grains smaller than a critical radius to the larger grains by several different diffusive mechanisms (Figure 3.16; see review in Hunter, 1996). If there is no liquid present then diffusion may be within the body of the grains or, more likely, along grain boundaries, both of which are slow. If a liquid is present then diffusion in the liquid dominates any within-grain diffusion. The latter is the most important mechanism for igneous rocks. The term *grain-boundary migration* is frequently used in igneous and meta-morphic petrology: it refers to diffusion either along grain boundaries or over a short distance via an intergranular liquid.

3.2.4.2 Lifshitz–Slyozov–Wagner coarsening model

Although the overall thermodynamic driving forces of coarsening are well known and understood the kinetics of the process are complex and the subject of much research, especially in the field of materials science (Voorhees, 1992). Unfortunately, the number of variables is very large and most of the systems

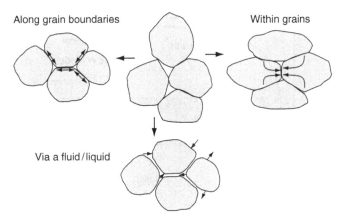

Along grain boundaries   Within grains

Via a fluid/liquid

Figure 3.16 Textural changes associated with diffusive creep involving different pathways for a system under a generalised vertical stress. Diffusion can be along grain boundaries (Coble creep); within the grains (Nabarro–Herring creep); or via an intergranular fluid/liquid. After Hunter (1996).

investigated are not appropriate to geological conditions. Nevertheless, a good starting point is the earliest and most well-known solution, that formulated by Lifshitz, Slyozov (1961) and others, known as the Lifshitz–Slyozov–Wagner (LSW) theory. The solution is for dilute systems, where there is assumed to be no interaction between crystals: all communicate directly and instantaneously with a uniform fluid. The crystals are assumed to be spherical. Most applications of LSW theory (and modifications of the theory) have been concerned with the relationship between mean crystal size and time. However, in plutonic rocks solidification time is not well constrained. Under these circumstances an equation relating growth rate to crystal size is more pertinent (Lifshitz & Slyozov, 1961).

$$\left(\frac{\mathrm{d}r}{\mathrm{d}t}\right) = \frac{k}{r}\left(\frac{1}{r^*} - \frac{1}{r}\right)$$

where $\mathrm{d}r/\mathrm{d}t$ is the growth rate, r^* is the critical radius, r is the crystal radius and k is a rate constant.

A graph of growth rates against radius shows that crystals below the critical radius are strongly resorbed and their material transferred to crystals slightly larger than the critical radius (Figure 3.17). The growth rate of large crystals tends to zero. The rate constant k is dependent on temperature and its value is not known.

Simple modelling of the effects of LSW coarsening on an initially straight CSD are shown in Figure 3.18a. There is a progressive loss of very small

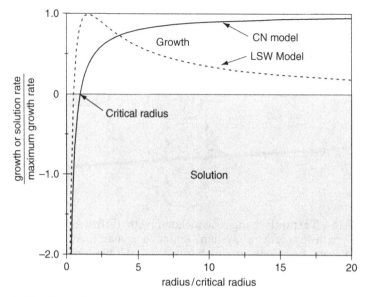

Figure 3.17   Growth rate variations for the LSW model and the Communicating Neighbours (CN) model (Higgins, 1998).

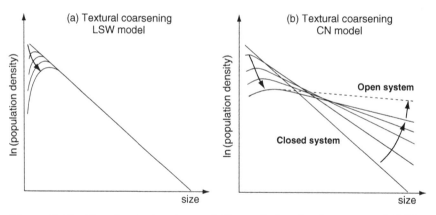

Figure 3.18   Simple stochastic modelling of the development of CSDs during LSW and CN coarsening (Higgins 1998). (a) For LSW coarsening the critical radius increases with each time step, but the rate constant is unchanged. (b) For CN modelling both the critical radius and the rate constant increase with each time step. Higgins (1998) proposed that open-system coarsening could reduce the slope of the CSD yet further.

crystals, but the actual numbers of larger crystals do not change very much. It is frequently emphasised that LSW Ostwald ripening will ultimately produce a CSD normalised to mean grain size that has a shape which is only dependent on the rate-controlling growth mechanism (Lifshitz & Slyozov, 1961).

However, this is probably not very useful in geological situations as complete equilibration needs constant conditions for a long time and is hence rarely achieved. Another problem is that the quality of much CSD data, especially for small sizes, is not sufficient to distinguish the different mechanisms.

3.2.4.3 Communicating Neighbours coarsening model

The limitations of the LSW equation are well known; hence it is not surprising that it is unable to model the observed CSDs. Numerous modifications have been proposed, such as extensions to high volume fractions (see review by Voorhees, 1992). However, as DeHoff (1991) has pointed out

'The literature abounds with theoretical treatments of coarsening for an additional important reason: the existing theories do not work.'

DeHoff proposed a new theory that he called 'Communicating Neighbours' (CN), that was applied to plutonic rocks by Higgins (1998).

CN theory is based on the observation that in many laboratory studies the growth rate of each crystal is not solely controlled by its size, but appears to be characteristic of each crystal, an effect called growth-rate dispersion. This suggests that the position of each crystal is important during growth and hence that crystals communicate with each other and not only with some uniform fluid. Of course, the mechanism for communication as in the LSW theory is by diffusion. DeHoff (1991) has formulated a geometrically general theory using this principle. The most readily applicable equation relates growth rate and crystal radius.

$$\left(\frac{dr}{dt}\right) = k\left\langle\frac{1}{\lambda}\right\rangle\left(\frac{1}{r^*} - \frac{1}{r}\right)$$

where dr/dt is the growth rate, k is a rate constant (different from that in the LSW theory), $\langle 1/\lambda \rangle$ is the harmonic mean of the intercrystal distance λ, r^* is the critical radius and r is the crystal radius. This equation illustrates the fundamental difference between the LSW and CN theories: in the former the diffusion length scale is a property of the particle itself, hence the $1/r$ dependence, whereas in the CN theory it is dependent on the distances to the neighbouring particles. This equation is for crystal growth controlled by diffusion. A similar equation also applies to interface-controlled growth (DeHoff, 1984).

A graph of growth rate against radius for the CN equation shows that crystals below the critical radius are resorbed rapidly, although less so than with the LSW equation. The big difference is for crystals larger than the critical radius, where the growth rate increases asymptotically with radius

(Figure 3.17). This equation is much more interesting than the LSW equation as the growth rate is constant for crystals much larger than the critical radius, and will then depend on undercooling, amongst other factors.

Higgins (1998) also applied stochastic modelling to the CN equation (Figure 3.18). The critical radius and rate constant were increased with equilibration to achieve the best fit to CSD data from the Kiglapait intrusion. The match for the right side of the CSD diagram is much better than the LSW model, although the turn-downs at small lengths are not as steep as observed. It should be noted that the common practice of plotting linear crystal population densities against length conceals the variations in the number of large crystals (e.g. Cashman & Ferry, 1988). Hence, the difference between the CN and LSW theories is not so clear on these diagrams.

3.2.4.4 Open systems and fluid focusing

The simplest models of coarsening assume that the system is closed. That is, all material that is dissolved from the small crystals will be precipitated on the larger crystals and hence the volumetric proportion of the phase is constant. However, the transfer of material is via a liquid phase and this phase can migrate into or out of a rock. Hence, the process of coarsening may be an open process.

Open-system processes may lead to local increases in the volumetric abundance of a phase. For example, in a crystal mush coarsening enhances the permeability by removing small crystals that may block channels. Increased permeability will, in turn, lead to focusing of fluid flow, further growth of megacrysts along the channels and positive feedback (Higgins, 1999). The channels formed by these interactions will appear as linear regions and nests of crystals in the two-dimensional surface of the outcrop (Figure 3.19).

3.2.4.5 Adcumulus growth

Many authors consider adcumulus growth to be an important process in the solidification of plutonic rocks, but the exact process has never been clearly defined. It is generally assumed that all crystals grow at the same rate (Figure 3.20). Hunter (1996) discouraged the use of this term because he felt that it was a model dependent term and not a descriptive term. He considered that adcumulus growth could not be distinguished from compaction. I propose that there is also a large component of coarsening involved, perhaps combined in some rocks by pressure solution compaction (Higgins, 2002b). Removal of small crystals ensures that permeability is maintained even at low degrees of porosity, hence allowing the development of rocks with little or no trapped liquid.

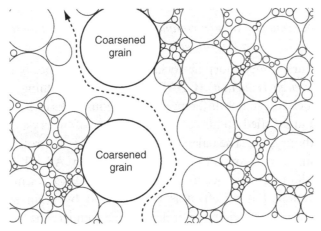

Figure 3.19   Open-system coarsening in crystal mushes. Coarsening removes small crystals and focuses flow. Very coarsened grains (megacrysts) form along the channels.

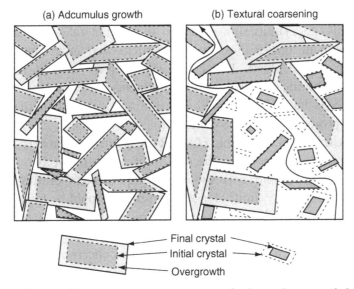

Figure 3.20   (a) Although the exact process of adcumulus growth has not been specified it is usually assumed that all crystals grow equally. (b) During coarsening small crystals dissolve and large crystals grow simultaneously. Permeability is maintained even at low porosity as small crystals that tend to block circulation are removed.

3.2.4.6 Coarsening in almost completely solid materials: metamorphic and sub-liquidus plutonic rocks

Coarsening occurs in metamorphic rocks where it is commonly referred to as 'grain-boundary area reduction' (GBAR) or Ostwald ripening. In deformed metamorphic rocks the major difference from igneous rocks is that excess energy is derived from unsatisfied bonds both at the surface of the crystals and around defects within the crystals. The balance of these factors is controlled by the grain size distribution, grain shape and defect concentration. All will be reduced as the texture equilibrates (see Sections 3.2.5 and 4.2.2). Coarsening can occur during deformation, but its effect is counteracted by grain size reduction processes. GBAR is most evident in rocks that have undergone '*static recrystallisation*', generally after deformation has ceased. If the process is to produce measurable effects then a fluid is necessary: one dominated by water at lower temperatures and by silicate at higher temperatures.

Coarsening of crystals suspended in a liquid, or free to move, is clearly very different from coarsening in a material that is almost completely solid: the presence of an independent second phase (or phases) may 'pin' the position of the grain boundary and limit coarsening in that direction (Berger, 2004, Berger & Herwegh, 2004). Such a situation occurs in all polymineralic rocks and may be important even in almost monomineralic rocks that only contain a small amount of a second phase. This effect is sometimes referred to as Zener-influenced coarsening. Modelling of the influence of second phases suggest that such CSDs have a lognormal distribution, rather than the Gaussian distribution produced by the LSW models.

The physical processes governing GBAR are identical to those of coarsening described previously. However, the saturation temperature of a phase in a metamorphic rock does not correspond to its liquidus temperature in magma, but to its upper stability limit. Hence, coarsening should be most important at higher temperatures, close to the stability limit of the phase. In metamorphic rocks kinetic effects must always be considered, particularly where fluid is not abundant. Hence, there has been much discussion concerning the applicability of coarsening and GBAR (Miyazaki, 1996, Carlson, 1999, Miyazaki, 2000).

3.2.5 Crystal deformation and fragmentation

3.2.5.1 Crystal deformation

Solid-state mechanical deformation of crystals and subsequent recovery of strain energy can result in grain size reduction (e.g. Karato & Wenk, 2002). This process differs from cataclasis in that at no point are the crystals mechanically broken. Crystals deform in the solid state by the production and migration of lattice defects.

Lattice dislocations are places in a crystal lattice where there is a mismatch between lattice planes (Figure 3.21). In an edge dislocation one lattice plane terminates and the space is adjusted by irregularities in the lattice. The atoms at the end of the lattice plane have unsatisfied bonds. In a screw dislocation the mismatch is restricted to a single point and the lattice is rotated helically about this point. Another type of lattice defect is a vacancy: here an atom is missing and the surrounding chemical bonds are unsatisfied. In both cases there is excess energy associated with the lattice defect.

Lattice dislocations have an unsatisfied bond that can be passed along to adjacent atoms readily in response to stress of the lattice (Figure 3.22). The

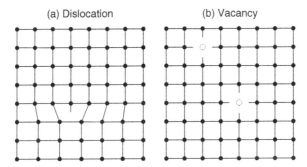

Figure 3.21   Lattice defects in crystals. (a) Dislocations are places within a crystal where a lattice plane terminates. The mismatch is accommodated by an irregular unit cell or cells. (b) Vacancies are lattice positions that are unoccupied.

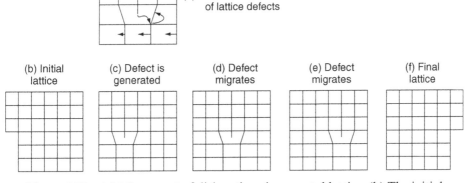

Figure 3.22   (a) Movement of dislocations in a crystal lattice. (b) The initial lattice has a step on the left. (c–e) A lattice dislocation can move across the lattice propagating a change in the shape of the crystal. (f) The final lattice has a step on the right.

overall effect of the passage of a dislocation through a lattice is that the crystal will change shape. Clearly edge dislocations are much more efficient than screw dislocations as a whole part of the lattice can be displaced. Dislocations can move along a number of different lattice planes. If several dislocations become entangled then movement may cease. This is the cause of strain hardening in materials. Such tangles can be removed if vacancies are allowed to jump over the entanglement. This effect is enhanced at higher temperatures.

The unfulfilled chemical bonds associated with lattice defects have a higher energy than other bonds in the lattice. Hence, the energy of a crystal can be minimised by a reduction in the number density of the defects. This is termed recovery. If the defects migrate to the edge of the crystal then the grain size is unchanged. However, it is generally more energetically favourable to accumulate the defects along surfaces within the crystals and hence form new, smaller subgrains. Each subgrain will have a slightly different orientation from its neighbours. The overall result of deformation and recovery is the subdivision of larger grains into a mosaic of smaller grains. There is some evidence that the CSDs of grain size reduced rocks may be fractal, that is they have a power-law distribution (e.g. Armienti & Tarquini, 2002).

A much more limited amount of deformation can be accommodated by twinning (Karato & Wenk, 2002). This effect is commonly seen in plagioclase and calcite, but can occur in a number of other minerals. Crystals can also deform by migration of lattice vacancies. This process is more important at higher temperatures.

3.2.5.2 Crystal fragmentation (cataclasis)

Fragmentation or cataclasis is the mechanical breakage of crystals in response to strain. It occurs if a crystal cannot deform fast enough by the production and migration of lattice defects. Hence, it occurs in rocks that are strained rapidly, or at lower temperatures. The exact threshold values for rapid strain are variable: some minerals, such as calcite are very weak and plastically deform easily. Other minerals, such as diamond, are strong, but may be broken if the emplacement process is sufficiently violent.

Strain can have an external or internal origin. External strain results from the deformation of the matrix, most commonly by other solid phases that are in contact with the grain, but it is also possible that rapid deformation of a very viscous liquid could induce fragmentation. Internal strain comes from changes in the volume of different parts of the crystal in response to changes in pressure or temperature. The most common source of internal strain is the effect of pressure on large fluid inclusions. The liquid and/or gas in inclusions has a very

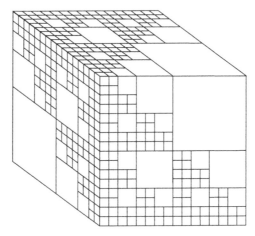

Figure 3.23   A fractal model for fragmentation after Turcotte (1992).

different compressibility to that of the crystal and hence pressure changes can generate large internal strains (Allen & McPhie, 2003).

One of the earliest models of rock fragmentation was developed by Kolmogorov (1941) and Epstein (1947). They proposed that breakage is a series of discrete events, and that during these events the probability of breakage of any clast is constant and independent of size. This model produces a lognormal size distribution by length. However, industrial crushing and grinding commonly produces fragments with the Rosin–Rammler distribution (Kotov & Berendsen, 2002). This results when fracture is controlled by initial flaws that have a Poisson distribution. This distribution is difficult to test and hence is transformed to the Weibull distribution (Kotov & Berendsen, 2002). Finally, the most recent fragmentation model is the fractal distribution.

The mathematical concept of fractional dimensions – fractals – is recent (Mandelbrot, 1982), but has had a remarkable impact on the natural sciences, particularly in geology (Turcotte, 1992). Indeed, it is now commonly assumed that most geological structures, including rock textures, are fundamentally fractal. A simple model of fracturing can give a fractal size distribution (Figure 3.23). Two diagonally opposite cubes are retained at each scale, and the other blocks divided into blocks with half the linear dimension. The fractal size dimension is 2.58 (Turcotte, 1992).

3.2.6 Closure in grain size distributions

The crystal content of a rock cannot exceed 100%; hence, as with chemical analyses, we must always be aware of the closure problem. Here the problem is treated for crystal size distributions, but a similar approach can be used for grain size distributions. Closure for an individual phase can also occur at less than 100% crystals. For instance, if a rock is made of 50% plagioclase and 50% olivine then closed-system processes, such as closed coarsening, cannot change the phase proportions – they are each fixed at a maximum of 50%. This effect must be considered in all CSD studies, both of volcanic and plutonic rocks (Higgins, 2002a).

We are concerned here with the effects of constant phase proportion on CSDs; however, Higgins (2002a) showed that even quite large variations in volumetric phase proportions can show the same effects. Clearly, closure is just a special case of this more general problem, where the volumetric phase proportion is equal to one.

For a population of crystals with constant shape, the volumetric proportion of phase i, V_i, can be calculated by integration of the volume of all the crystals, following Higgins (2002a):

$$V_i = \sigma \int_0^\infty n'_{V_i}(L) L^3 \mathrm{d}L$$

where σ = shape factor of phase i, $n'_{V_i}(L)$ = population density of crystals of phase i for size L. The shape factor is equal to the ratio of the actual volume of the grain divided by the volume of a cube that encloses the grain, L^3. It can be expanded into a more applicable form as follows

$$\sigma = (1 - \Omega(1 - \pi/6)) \mathrm{IS}/L^2$$

where Ω is the roundness factor, which varies from 0 for rectangular parallelepipeds to 1 for a triaxial ellipsoid. S, I and L are the short, intermediate and long dimensions of the parallelepiped or ellipsoid. Obviously most crystals have a more complex shape, but we are concerned here just with a statistical measure. However, the above equation should not be applied to crystals with concave surfaces or holes, as they do not necessarily have volumes that lie between that of an ellipsoid and an equivalent parallelepiped.

For straight CSDs (semi-logarithmic) with intercept $n'_{V_i}(0)$ and characteristic length C ($C = -1/\text{slope}$) a simplified equation can be used

$$V_i = 6\sigma n'_{V_i}(0) C_i^4$$

It should be emphasised that volumetric phase proportions, V_i, calculated using these equations are sensitive to errors in the shape factor – it is common to find errors up to a factor of two. Hence the actual phase volume should be determined from the total area of the crystals (see Section 2.7).

The correlation between the slope and intercept of straight CSDs has been noted by many authors (e.g. Higgins, 1999, Higgins, 2002a, Zieg & Marsh, 2002). Zieg and Marsh (2002) linked slope and intercept by a 'modal normalisation factor' which incorporates the inverse of the modal abundance and a shape factor. They attribute this relationship to scaling laws, but it is clear that it is just a geometric relationship referred to here as closure.

At low volumetric phase proportion, for example 1%, the slope of a straight CSD can change independently of the intercept (Figure 3.24a). For example, if all crystals grow at the same rate and nucleation increases exponentially, then the CSD will move up without changing slope. Once the crystal content reaches 100% then closure is reached and the CSD is locked. It can only move if the ratio between crystals of different sizes is changed. That is, some crystals must become smaller if others are to enlarge.

A family of straight CSDs for 100% crystallised material defines a fan (Figure 3.24b; Higgins, 2002a, Zieg & Marsh, 2002). Portions of this fan appear to describe rotation of the CSD around a point commonly close to the left of the diagram. The CSDs together outline a concave-up envelope.

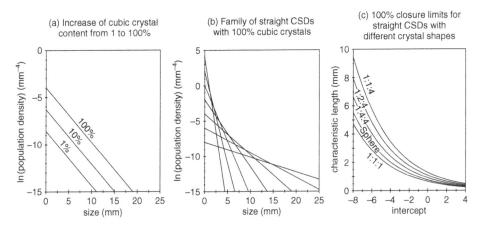

Figure 3.24   (a) The closure problem in CSDs, illustrated for cubic grains (Higgins, 2002a). A straight CSD with a slope of −0.57 (characteristic length 1.73 mm) and an intercept of −8.6 has a volumetric phase proportion of 1%. If the intercept is increased to −4.0, for example by crystal growth and exponential increase in nucleation rate, then the crystal content will reach 100% and textural changes will stop. (b) A fan of straight-line CSDs with 100% crystals is tangential to a concave-up curve. (c) Straight CSDs can be represented by a point on a graph of characteristic length (−1/slope) against intercept. Slope and intercept values for materials with 100% crystals proscribe a curve. CSDs can exist below, but not above this line. The position of the line differs for different shapes, here indicated by the aspect ratios of rectangular parallelepipeds and spheres. It is assumed here that the crystals completely fill the volume.

Figure 3.24c shows another view of the same effect in terms of characteristic length against intercept. All straight CSDs can be defined by their intercept and characteristic length. Closure limits describe a curve for each crystal shape. Straight CSDs can only exist below this line.

If a sample has a low crystal content, that is it is far from the closure limit, then the intercept and slope can change independently giving two degrees of freedom. However, as the crystal content approaches 100% then any process that changes the slope of the CSD must also change the intercept – in this situation there is essentially only one degree of freedom. Also, if the crystal content is fixed (e.g. $V_i = 10\%$, 50%, 100% etc.) then intercept and slope must be also linked. Therefore, crystals must dissolve (melt) or divide into sub-grains to enable changes in the slope and intercept of the CSD.

3.3 Analytical methods

Grain sizes and size distributions are three-dimensional properties and ideally they are measured directly. However, as was mentioned in Chapter 2, in many materials this is either not possible or not the best approach. In those cases data must be acquired from two-dimensional sections and the true sizes calculated using stereological methods. Whatever method is used, it is important that researchers clearly describe their method and make publicly available the basic data of a study in a useable format. In this way data can be recalculated and comparison between new and existing studies is made possible.

It is also important to decide and state what material is being measured. This is particularly important for rocks in which some or all of the crystals are broken. If broken crystals are included in the CSD then the number of larger crystals will be reduced and that of smaller crystals increased. The effect will be much more significant for small crystals if their actual abundance is low or zero. It is not possible to reconstruct the crystal CSD from a mixed population of crystals and fragments. A possible solution is to examine each crystal individually and determine if it is a fragment. If only a few crystals are broken then a CSD of intact crystals may be determined. If most of the crystals are broken then there is nothing to be done.

3.3.1 Three-dimensional analytical methods

3.3.1.1 Solid materials

The size of individual grains can be determined by serial sectioning (see Section 2.2.1). In transparent materials, such as volcanic glasses, the size of crystals can be measured directly with a regular or confocal microscope (see

Section 2.5.3). X-ray tomography can be used to determine the size of grains if adjoining grains can be separated (see Section 2.2.3). The size of crystals is preserved if they can be extracted intact from rocks by sample disintegration and solution (see Section 2.4). It is then possible to use techniques developed for unconsolidated materials (see below).

3.3.1.2 Unconsolidated or disaggregated materials

The size distribution of loose materials may be determined directly by a number of methods, some of which use unique definitions of size (see below; Syvitski *et al.*, 1991). The most direct method of grain size analysis is sieving. Sieves are available in a number of standard sizes and materials. Disposable nylon sieves eliminate cleaning, which can be laborious. Sieves are not very efficient for grains less than 500 μm. Ideally the longest axis of the grain passes vertically through the sieve hole. Under these conditions the maximum size of grain that can pass through a sieve is equal to an intermediate dimension of the grain.

The principal use of grain size analysis in sedimentology is to examine sedimentation patterns. These depend at the first approximation on the settling characteristics of particles. Hence, several analytical methods use controlled sedimentation to measure size (Syvitski *et al.*, 1991). Hence, grain size is defined by settling velocity and converted to equivalent sphere diameters. One of the more popular instruments uses X-rays to measure the amount of sediment that is deposited at the base of the settling tube (Coakley & Syvitski, 1991).

Another method uses changes in electrical resistance to determine the size and volume of grains (Milligan & Kranck, 1991). The sample is introduced into an electrolyte, mixed and then drawn through a narrow aperture. A constant current passes between the two electrodes on either side of the aperture. Each particle displaces an equivalent volume of the electrolyte and hence changes the resistance. If the particles are allowed to pass one at a time, then each particle will produce a pulse that is proportional in size to the volume of the particle.

Rapid measurements of grains smaller than 1 mm are possible with a laser particle diffraction spectrophotometer (Agrawal *et al.*, 1991). The sample is dispersed in water which is introduced into a transparent cell. Light from one or two lasers is shone through the cell and allowed to illuminate an array of concentric circular detectors. Small particles diffract the light more than larger particles. Hence, the variation of intensity versus deflection angle can be transformed into the grain size distribution. Measurement is instantaneous since there is no sedimentation effect.

3.3.2 Two-dimensional analytical methods

The size of grains can be determined from images acquired from rock surfaces, slabs, sections and projections, using techniques described in Section 2.2. Such images can be processed manually or automatically to produce classified mineral images (see Section 2.3). Finally, the binary images must be processed to extract crystal outline parameters, such as size.

Grains can be intersected by or projected onto a plane to give a grain outline (Figure 3.25). Thin sections have a finite thickness and hence small crystals may be projected whereas larger crystals are intersected. These situations must be clearly distinguished as the conversions to 3-D data are different. Grain outline definitions will be discussed first, followed by intersection and projection methods.

3.3.2.1 Grain outline width and length definitions

If a sphere is intersected by a plane, or projected onto one, then the outline shape is a circle. A circle has a unique, clearly defined length or width: its diameter. Crystal or bubble outlines generally have irregular shapes; hence outline width and length are not uniquely defined. The area of each outline is uniquely defined, but is not the best parameter for stereological calculations of size (Higgins, 1994). However, in a section the total area of grain intersections divided by the area measured is equal to the volumetric proportion of the phase (see Section 2.4.2; Delesse, 1847). Hence, the intersection area is significant and should be measured if possible.

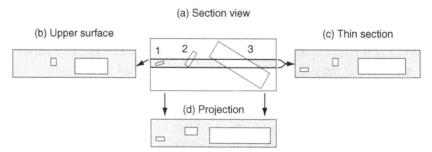

Figure 3.25   (a) Section view of three opaque crystals in a transparent matrix. A thin section is indicated in pale grey. (b) The upper surface of the thin section is imaged. Crystal 1 is not visible. (c) The crystals are imaged in a thin section. All three crystals are imaged. The outline of crystal 3 is larger than that on the upper surface. (d) The crystals are viewed in projection. Crystal 1 presents the same outline as in the thin section. Crystals 2 and 3 have larger outlines than those in other views.

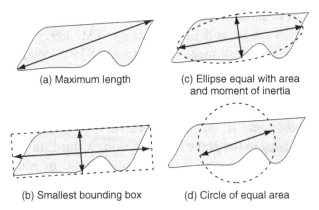

Figure 3.26   Commonly used definitions of outline size. (a) Greatest length. (b) Length and width of smallest bounding box. (c) Major and minor axes of an ellipse of equal area and moment of inertia. (d) Circle of equal area.

Several definitions of outline length and width are possible (Figure 3.26).

1. The length can also be defined as the greatest distance between points on the outline. This is generally greater than the 'box length'. This is sometimes known as the Maximum Feret length.
2. A box can be rotated around the outline until the box width is narrowest or the box length is greatest – these are commonly the same orientation. The outline length and width are the dimensions of this box. This is used by some stereological corrections programs (e.g. CSDCorrections).
3. The outline can be approximated by an ellipse of equal area and moment of inertia. The outline length and width are the major and minor axes of the ellipse. This is used by the popular free program NIHImage (see Appendix).
4. The outline can be modelled by a circle of equal area. The outline size is the diameter of this circle and is commonly called the Feret diameter. This is not a good choice as it compresses the sizes of outlines.
5. Some programs calculate the width and height of a bounding box with vertical and horizontal sides (e.g. NIHImage). Clearly this 'size' depends on the orientation of the crystal outline and is not useful.

3.3.3 Intersection mean size methods

3.3.3.1 ASTM mean and maximum size methods

The American Society for Testing Materials (ASTM) has developed standard procedures for the determination of the apparent grain size of metals from grain intersection data. The standards only assess apparent mean grain size in sections and do not make any attempt to extend this to the true 3-D grain size.

The first standard deals with the problem of estimating the largest grain in a section, called ALA (as large as) grain size (ASTM, 1992). A further, more detailed, standard deals with mean grain sizes (ASTM, 1996). The standard suggests two analytical methods: number of grains per unit area and number of grain intersections per unit length of test line, with the emphasis on the latter method. If a metal has a significant orientation fabric then the measurement of three orthogonal sections is advocated. Circular test lines also compensate for anisotropy in the plane of the section. A third standard deals with mean grain size measurements using semi-automatic or automatic image analysis methods (ASTM, 1997). Measurements of 400 grains give an uncertainty of about 10% in the mean grain size. It should be remembered that these procedures do not give grain size distributions, just maximum and mean grain sizes. Apparent grain sizes are expressed as the ASTM grain size number, G, where N_A is the number of grains per square mm, and N_L is the number of grain intercepts (times entering a grain) per unit length of test line:

$$G = 3.32 \log(N_A) - 2.95$$

$$G = 6.64 \log(N_L) - 3.29$$

3.3.3.2 Grain number by shape independent intersection methods

There are a series of methods that use two parallel sections to determine the total number of grains per unit volume, N_V (Royet, 1991). The advantage of these methods is that the shape and size of the grains is not important. The basic method is called the 'Dissector'.

A 'reference section' of area A is defined and grain outlines are determined. The 'lookup section' is parallel to the reference section and separated by distance H. Grains whose outlines are present in the reference section are looked for in the lookup section. Q is the number of outlines counted in the reference section that cannot be identified in the lookup section. Then:

$$N_V = Q/(AH)$$

The resolution of the method is limited by the spacing of the reference and lookup sections, and by the ability to trace grains from one section to the other. It should be emphasised that this method gives the total number of grains per unit volume and not the grain size distribution. The mean volume of the grains can be determined if the total volume of the grains is known (see Section 2.4.2). The mean size of the grains can be calculated if a grain shape is assumed.

3.3.3.3 Mean grain size by shape independent intersection methods

The ratio of the surface area to the volume, S_V, can be accurately determined without any assumptions of grain shape or orientation (see Section 2.4.2). This parameter is a measure of the mean curvature of the grains and has the dimensions of L^{-1}. The inverse of this parameter is a measure of the mean size of the grains, which is just the mean intercept length. This measure of mean size is used in some materials science literature (Brandon & Kaplan, 1999).

3.3.4 Extraction of grain size distributions from grain intersection data

3.3.4.1 Nature of the problem

The problem of conversion of two-dimensional intersection or projection size data to three-dimensional CSDs is not simple for objects more geometrically complex than a sphere and there is no unique solution for real data (Figure 3.27).

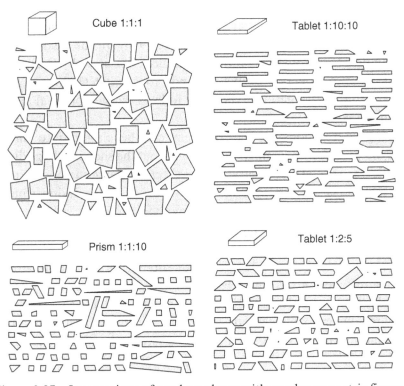

Figure 3.27   Intersections of random planes with regular geometric figures. It should be noted that tablets tend to give elongated intersections, which are commonly thought to indicate a lath shape, whereas prisms commonly have a rectangular cross-section (Higgins, 1994).

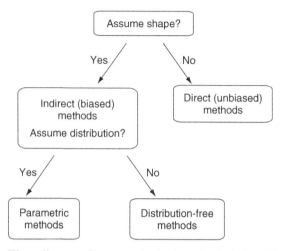

Figure 3.28   Flow diagram for stereological methods (after Higgins, 2000).

The problem was first attacked in geology by sedimentologists: empirical solutions were developed early and, despite criticism, have been commonly applied since then (Friedman, 1958, Johnson, 1994). In the field of igneous and metamorphic petrology the problem has been discussed by Cashman and Marsh (1988), Peterson(1996), Sahagian and Proussevitch (1998) and Higgins (2000). The following treatment generally follows the stereological approach of Higgins (2000).

Stereological solutions to this problem can be classed into *direct* or *unbiased* methods for which no assumptions about the shape and size distribution of the particles are necessary, and *indirect* or *biased* methods, in which some assumptions are needed (Figure 3.28; Royet, 1991). Indirect methods can be classified further into *parametric* methods, where a size distribution law is assumed, and *distribution-free* methods that are applicable to all distributions.

In an ideal world the direct methods would be the best solution. However, although direct methods for the determination of mean size are very powerful and simple (Howard & Reed, 1998), they are not available for size distributions. Three-dimensional methods can yield size distributions directly, but their limitations have already been discussed (see Section 2.2). The lack of a direct size distribution method has prompted some workers in stereology to suggest that size distributions should not be determined from sections, or that size distributions do not give any useful information to practical problems (Exner, 2004). These limitations, however, seem unduly onerous and hence we are usually left, therefore, with indirect methods.

It should be mentioned that the techniques to be discussed below are based on the assumption that the section measured (or visible part of the surface) is thin compared with the dimensions of the object. For thin sections viewed in transmitted light, this assumption means that the crystals must be larger than 0.03 mm, the thickness of a standard thin section. If crystals smaller than this limit are to be measured, then the crystal outlines are a projection and not a section, and hence different equations must be used (see Figure 3.25 and Section 3.3.5).

3.3.4.2 Parametric solutions

The earliest geological methods for stereological conversion were determined from studies of loose sediments, supplemented by determinations of grain size from thin sections. However, it was discovered early that there was a systematic bias in thin section data as compared to sieved fractions (Chayes, 1950). The earliest attempts to solve this problem were based on empirical data: sediments were analysed by sieving and thin-section analysis (Friedman, 1958). True grain size distributions are commonly approximately lognormal by mass at the precision level of this type of analysis; hence this method is actually parametric. The core of the method is to use the quartiles from the analysis of size in thin sections and transform these to quartiles equivalent to those of the sieved data using a simple linear equation. This method has been corrected and modified by others, but was most severely criticised by Johnson (1994). He pointed out that the method does not work very well for the smaller size fractions, but only suggested a way of calculating the mean particle size from intersection measurements.

Kong *et al.* (2005) proposed another parametric solution for spheres with lognormal or gamma distributions. Despite the complexity of the mathematical arguments, this method is very restrictive in terms of shape and distribution: it will be shown below that there are much simpler and more powerful solutions to this problem. They also concluded that a simple factor can be used to convert 2-D mean sizes to 3-D mean sizes. Again, a much better solution is to use stereologically exact global parameters (see Section 2.5.2) as there is no need to assume shape or distribution.

Peterson (1996) proposed a parametric solution based on another distribution: he initially assumed that rock CSDs have a strict logarithmic variation in population density for a linear variation in size. This distribution is indicated by the theoretical studies of Marsh (1988b) for simple igneous systems. Peterson first corrected the data for the intersection-probability effect (see below). He then applied a further correction using three empirical parameters that depend on the shape and spatial orientation of the crystals. He found a

good correlation between the CSDs determined by his methods and the true
CSDs for synthetic data with logarithmic-linear CSDs. However, many
published natural CSDs are not linear. Peterson was aware of this problem
and applied the corrections derived from linear approximations to the
actual curved CSDs. However, he did not verify whether the results for
curved CSDs were accurate using synthetic or natural data. Hence, it is
not clear if these parametric methods can be applied to many natural
systems. It also seems unwise to use a technique to find CSDs that assumes a
CSD shape beforehand. Another limitation of these methods is that they can
only be applied to isotropic (massive) fabrics, unless other parameters are
introduced.

3.3.4.3 Distribution-free solutions

The intersection-probability and cut-section effects are relevant for distribution-
free solutions (Underwood, 1970, Royet, 1991). The intersection-probability
effect is that for a population of grains with different true sizes (polydisperse),
smaller grains are less likely to be intersected by a plane than larger grains. The
cut-section effect is that the intersection plane never cuts exactly through the
centre of each particle; hence, even in a population of equal-sized particles
(monodisperse), intersection sizes have a broad range about the modal value,
from zero to the greatest 3-D length. Both problems compound to make the
stereological conversion complex.

The intersection-probability effect is easily resolved for a monodisperse
collection of spheres (Royet, 1991):

$$n_V = \frac{n_A}{D}$$

This equation can be modified so that it applies to polydisperse (many true
sizes) collections of spheres if many size ranges are treated separately: for
instance 0–1 mm, 1–2 mm, 2–3 mm. Other shapes can be accommodated if a
linear measure of the size of the particle is used instead of the diameter.

The simplest formulation of the intersection-probability effect for non-
spherical objects in a size interval j is:

$$n_V(L_j) = \frac{n_A(l_j)}{\bar{H}_j}$$

Mean projected height (or mean calliper diameter) $\bar{H}_j$ is defined as the mean
height of the shadow of the particle for all possible orientations (Sahagian &
Proussevitch, 1998). It is commonly close to the other definitions of size
discussed in Section 3.1.2.

Table 3.1   *Some symbols that are used below. Lower case letters*
commonly refer to intersections and upper case to 3-D parameters.
Other symbols are discussed later.

j	A size interval
x, y	Size limits of an interval
L	3-D size
W_j	Width of interval j
L_j	Mean size in interval j
D	Sphere diameter
V	Volumetric phase proportion
l	Intersection length
l_j	Mean intersection length in interval j
w	Intersection width
w_j	Mean intersection width in interval j
P_{AB}	Probability that a crystal with a true size in size interval A will have an intersection that falls in intersection size interval B
$\bar{H}_j$	Mean projected height for size interval j
n_A	Total number per unit area
n_V	Total number per unit volume
$n_A(l_j)$	Number per unit area for size interval j
$n_V(L_j)$	Number per unit volume for size interval j
$n_V(0)$	Nucleation density (intercept on size axis)
$n'_V(L)$	Population density at length L
Shape parameters	
S	Short axis of parallelepiped or minor axis of ellipsoid
I	Intermediate axis of parallelepiped or intermediate axis of ellipsoid
L	Long axis of parallelepiped or major axis of ellipsoid

The intersection probability effect uses the assumption that each intersection passes exactly through the centre of the object along the longest axis. This is very unlikely and hence we have to contend also with the cut-section effect.

The cut-section effect can be solved analytically only for spheres (e.g., Royet, 1991). The size distributions of intersections between a random plane and a sphere of unit diameter rises to a maximum near the diameter of the sphere (Figure 3.29a). Hence, the most likely intersection diameter is close to the maximum and the true diameter.

The intersection of a triaxial ellipsoid with a plane produces an ellipse. The distribution of the major and minor axes (= length and width) of the intersection can be evaluated numerically (Figure 3.29). The intersection lengths of oblate ellipsoids (Figure 3.29a) and intersection widths of prolate ellipsoids (Figure 3.29d) have size distributions that resemble spheres. Intersection lengths of prolate ellipsoids (Figure 3.29b) also rise to a maximum at the

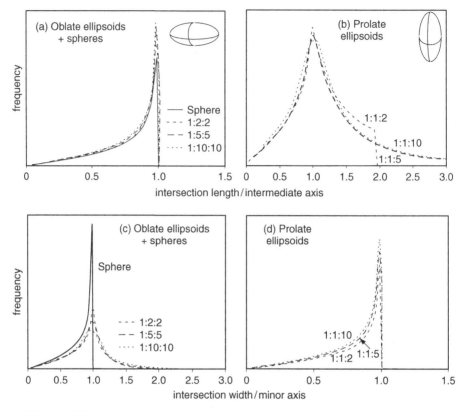

Figure 3.29   Intersections between a random plane and a triaxial ellipsoid (this study). One million randomly oriented intersections were calculated for different shapes of ellipsoids. Intersection lengths and widths are normalised to true (3-D) intermediate and minor axes of the ellipsoid. Oblate ellipsoids are short and wide and prolate ellipsoids are tall and thin.

intermediate axis value, but continue to larger sizes. Intersection widths of oblate ellipsoids rise to a maximum for the minor axis and also continue to larger sizes (Figure 3.29c).

The distribution of intersection dimensions for shapes such as parallelepipeds must be derived numerically and can be very complex with several modes (Figure 3.30; Higgins, 1994, Sahagian & Proussevitch, 1998).

As noted above, for spheres and near-equant forms, the mean intersection length is close to the true 3-D size of the object. However, for monodisperse populations of randomly oriented anisotropic figures, such as parallelepipeds, the most-likely intersection length is close to the intermediate dimension (Higgins, 1994). That is, for a particle $1 \times 2 \times 10$ mm, the most-likely intersection length is 2 mm. The most-likely intersection width is close to the short dimension. The same argument can also be applied to populations of crystals

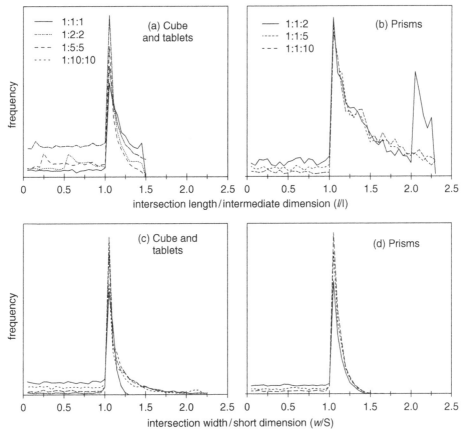

Figure 3.30 Intersections between a random plane and a regular parallelepiped. Intersection lengths and widths normalised to intermediate and short parallelepiped dimensions (Higgins, 1994, Higgins, 2000).

with different true sizes. These modal values for intersection length and width can be corrected to the true length of the crystal, or any other size parameter, if the shape of the solid is known (see Chapter 4). However, there is another problem: intersection lengths and widths for a monodisperse population tail out to smaller and larger values respectively around the most-likely intersection length. If the simple conversion equations are used, then the effects of tailing become very important. This is because small intersections of large objects (corners) will be converted using their apparent length and not their true length.

3.3.4.4 Saltikov method

Saltikov (1967) proposed a method[3] of unfolding a population of intersection lengths into the true length using a function of the intersection lengths (Figure 3.31). For a polydisperse population, the 3-D distribution of lengths

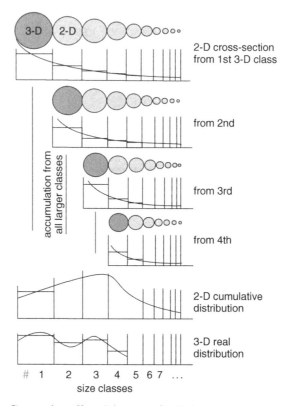

Figure 3.31   Cut-section effect. Diagram after Sahagian and Proussevitch (1998).

can be found by applying the function for a monodisperse population of the same shape to the 2-D length distribution. The values of $n_V(L_{XY})$ for a series of bins from 1, 2, 3, 4... can be calculated sequentially using the equations of Sahagian and Proussevitch (1998) as modified by Higgins (2000):

$$n_V(L_1) = \frac{n_A(l_1)}{P_{11}\bar{H}_1}$$

$$n_V(L_2) = \frac{n_A(l_2) - n_V(L_1)P_{12}\bar{H}_1}{P_{22}\bar{H}_2}$$

$$n_V(L_3) = \frac{n_A(l_3) - n_V(L_2)P_{23}\bar{H}_2 - n_V(L_1)P_{13}\bar{H}_1}{P_{33}\bar{H}_3}$$

$$n_V(L_4) = \frac{n_A(l_4) - n_V(L_3)P_{34}\bar{H}_3 - n_V(L_2)P_{24}\bar{H}_2 - n_V(L_1)P_{14}\bar{H}_1}{P_{44}\bar{H}_4}$$

where subscript 1 refers to the first (largest) size interval.

The size intervals (bins) are logarithmic as this simplifies the calculations. However, programs such as CSDCorrections can accommodate any size bins. Both Saltikov (1967) and Sahagian and Proussevitch (1998) proposed ten size bins per decade, that is each bin is $10^{-0.1}$ smaller than its neighbour. Higgins (2000) suggests 5 bins per decade as a starting value as this reduces the total number of bins and hence the accumulation of uncertainty. It also yields larger numbers of crystals, and smaller counting errors, in each bin where applied to typical geological samples. The probabilities, P_{AB}, can be calculated from numerical models of different crystal shapes (Higgins, 1994, Sahagian & Proussevitch, 1998). For isotropic fabrics, the shape is constructed mathematically and sectioned using randomly oriented planes placed random distances from the centre of the crystal model. The mean projected heights also can be calculated using the same models.

The Saltikov method works well for spheres and near equant objects (spheroids) because the modal 2-D length lies close to the maximum 3-D length (e.g., Armienti *et al.*, 1994, Sahagian & Proussevitch, 1998). However, Higgins (2000) has shown that this method is less successful for complex objects because the largest 2-D intersections are not very common. That is, the modal 2-D length is much less than the maximum 2-D length. Therefore, the less abundant, and hence less accurately determined, larger intersections are used to correct for the most abundant intersections, introducing large uncertainties in the corrected 3-D length distributions (Figure 3.32a). The problem is actually worse than might be expected from the works of Saltikov (1967) and

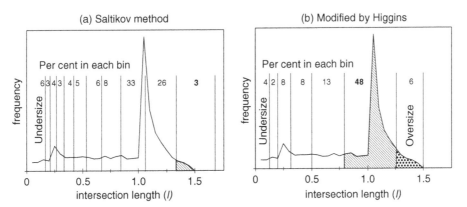

Figure 3.32 The Saltikov method for a tablet 1:5:5. (a) The Saltikov (1967) method uses the largest interval (shaded) for correcting the smaller intervals. For a 1:5:5 tablet this interval has only 3% of the intersections and is hence imprecise. (b) Higgins (2000) modified the Saltikov method by using a much wider bin that contains 48% of the intersections and is hence more precise.

Sahagian and Proussevitch (1998) as both authors placed the maximum inter-section length at the upper limit of the size bin. Because true particle sizes can be distributed throughout the size bin, at the very least the maximum inter-section size should be placed at the centre of the bin.

Sahagian and Proussevitch (1998) reduced this problem by ignoring the class of largest intersections and shifting the probabilities to lower classes. Higgins (2000) used the most likely intersection length or width (modal value) to correct for the tailing to other intersections (Figure 3.32b). However, it is not possible to correct tailing in both directions using this technique – tailing to intersections either larger or smaller than the modal value must be ignored. In general tailing to smaller intersections is a much more important problem than tailing to larger intersections for most CSDs. Hence, the first size interval is centred on the mode of the intersection length or width.

Some workers have considered the Saltikov technique to be impractical because of the accumulation of terms, and hence uncertainties, in the smaller bins. This problem is not always very serious as many of the terms are not significant, especially for equant and spherical forms. It can also be reduced by using fewer, wider bins. Intersections larger than the first interval cannot be corrected precisely with this method. However, if they are added to the first interval, then this source of uncertainty is minimised.

Clearly, the fewer corrections that are needed the greater the accuracy of the final data. In many cases use of intersection widths instead of lengths reduces the amount of corrections necessary (Figure 3.33).

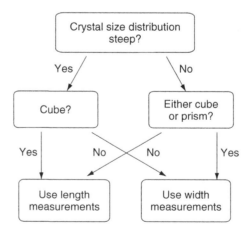

Figure 3.33 A possible strategy for choosing width or length intersection measurements (Higgins, 2000).

Once the mode of the intersection length or width is used instead of the maximum values, then the intersection length scale no longer corresponds to the true length scale: that is, $L_x \neq l_x$. For a parallelepiped with a short dimension S, an intermediate dimension I and a long dimension L, or an ellipsoid with the same dimensions, the modal value of the intersection lengths on randomly oriented planes is I (Higgins, 1994) hence:

$$L_x = l_x L / I$$

Similarly, the modal value of the intersection widths is S hence:

$$L_x = w_x L / S$$

Higgins (2000) found that for tablets, where $L/I = 1$, a much better fit is obtained to the test data of Peterson (1996) if the mean intersection length is used rather than L in the above equations. This gives a maximum difference of 20% in the length scale for 1:10:10 tablets and decreases to zero for cubes. The Saltikov equations do not need to be changed as $\bar{H}_1$ is determined in terms of the true crystal length.

Other refinements have also been made to the Saltikov method. Tuffen (1998) has pointed out that the inverse of the mean projected height must be averaged over the true length interval and proposed that:

$$\left(\frac{1}{\bar{H}}\right)_{xy} = \frac{1}{x - y} \int_{y}^{x} \frac{dL}{L}$$

where $x =$ upper limit of interval and $y =$ lower limit of interval. This can be integrated to give:

$$\left(\frac{1}{\bar{H}}\right)_{xy} = \frac{\ln(x/y)}{x - y}$$

The original equations of Saltikov (1967) used logarithmic size bins. Sahagian and Proussevitch (1998) followed this idea, but introduced more complex equations for linear size bins. However, if the probabilities used in the Saltikov equations are calculated for each cycle of corrections then the calculations remain simple for any set of bin sizes. This method of calculation has been implemented in the program CSDCorrections (Higgins, 2000), so that previously published data can be corrected. Nevertheless, the use of logarithmic size bins is recommended for most studies.

The Saltikov method can only be used in its simplest form for isotropic fabrics. However, the probabilities of intersection, and the factors for converting intersection widths and lengths to true crystal sizes can be readily modelled for anisotropic fabrics if the fabric parameters are known or can be estimated. This is done on a dynamic basis in the program CSDCorrections (Higgins, 2000). The intersection probabilities also depend on the shape of the crystals. Crystal shapes will be discussed in Chapter 4.

There have been a number of published methods which are modifications of the Saltikov method, although the authors were not always aware of this at the time. Pareschi *et al.* (1990) developed a stereological procedure that they called 'unfolding' which closely resembles the Saltikov method. In this method they reduced the intersection area to a circle of equal area, but otherwise used the Saltikov technique for spheres. The 'Stripstar' technique and program is again very similar (Heilbronner & Bruhn, 1998). Crystals are assumed to be spheres and simple corrections are made using a fixed number of fixed-width size intervals. The weaknesses of fixed-size intervals have been noted above. Variable-size intervals can be readily accommodated if population density instead of frequency is used in the final CSD diagrams (see Section 3.3.7). In this technique, if all intersections from larger spheres are stripped off from smaller size bins then negative quantities of spheres (antispheres) are permitted, which seems to be counterproductive. Programs such as CSDCorrections (Higgins, 2000) can do the same type of simple corrections using spheres, with a flexible choice in size intervals and without negative quantities of crystals.

3.3.4.5 Uncertainty analysis of the Saltikov method

Inaccuracy in the determination of population densities arises principally from three sources (Higgins, 2000): the easiest to understand and quantify is the counting uncertainty. This is taken to be the square root of the number of intersections within an interval. It is usually only significant for larger size intervals with fewer than 20 intersections.

The second source of uncertainty is in the value of the probability parameters P_{AB} used in the Saltikov equations. Although these parameters are defined precisely for fixed convex shapes, crystals in most natural systems have more irregular and variable shapes. Another source of uncertainty is that tailing to intersections larger than the modal interval is included in the modal interval. Hence, it is difficult to estimate the contributions from this source to the total uncertainty. However, it is easy to calculate the contribution of the counting uncertainties of other intersection intervals to the total correction of an interval. This source of uncertainty is most important for small size intervals, where corrections are most significant.

The third source of uncertainty lies in the conversion of intermediate crystal dimensions (for intersection length measurements) or short crystal dimensions (for intersection width measurements) to true crystal lengths. Uncertainties in the determination of the crystal shape will produce systematic uncertainties in both the population density and the size distribution.

3.3.4.6 Data conversion methods used in early CSD studies: Wager and Grey methods

Many early crystal and vesicle size distribution studies used the following equation to convert intersection size data to CSDs:

$$n_V(L_j) = n_A(l_j)^{1.5}$$

for each size interval *j*.

This equation has been ascribed to many different workers, but appears to have been first used by Wager (1961) to convert total numbers of crystals in sections to volumetric numbers, without any discussion of its origin or justification for its use. Wager (1961) may have chosen this equation because it is the simplest way to convert the dimensions of $n_A(l_j)$, $1/L^2$, to that of $n_V(L_j)$, $1/L^3$. This equation is not discussed in general reviews of stereological methods (e.g. Underwood, 1970, Royet, 1991, Howard & Reed, 1998) or those applied to geological problems (e.g. Sahagian & Proussevitch, 1998). It is not correct and the use of this equation should be discontinued.

Published data determined using Wager's equation are not wasted and can be recalculated if information on grain shape, grain fabric and orientation of the section to the fabric are available or can be estimated. The easiest way is to use tables of intersection data. If these are not available then the published CSD graphs can be measured and the size intervals and erroneous population density determined. In some graphs the actual or base 10 logarithm of the erroneous population density is plotted and this must be converted to the natural logarithm of the erroneous population density. The Wager equation is then applied in reverse to the erroneous population density of each size interval and the true value of N_A recovered. This can then be used to calculate the true N_V, using a stereologically correct method (see sections above).

Another method of converting intersection dimensions to three-dimensional data was proposed by Grey (1970). This equation can be applied to each size interval

$$n_V(L_j) = \frac{(\pi\alpha)^{1/2}n_A(l_j)}{4r}$$

where α is the aspect ratio (intersection length/intersection width ratio $= l/w$) and r is the radius of a circle of area equal to the area of the crystal outline. If the intersections are assumed to be square this equation can be recast as

$$n_V(L_j) = \frac{\sqrt{\pi l/w}}{4\sqrt{lw/\pi}} n_A(l_j)$$

This can be simplified to

$$n_V(L_j) = \frac{\pi\alpha}{4} \cdot \frac{1}{l} n_A(l_j)$$

In this form it is clearly similar to the classic equation $n_V(L_j) = n_A(l_j)/\bar{H}_j$, but crystal sizes will be offset by a factor of $\pi\alpha/4$ from the true values.

3.3.5 Extraction of grain size distributions from projected grain data

The stereological corrections for grains measured in projection are generally much simpler than those for intersections described above. However, it is essential that the whole object is seen – that is it does not emerge from either surface. The intersection probability effect does not exist; hence projective methods are equally sensitive for small and large grains. The cut-section effect is replaced by the projected outline effect. As before this can be assessed using two simple models: ellipsoid and parallelepiped. The most common use of projected images is in sedimentary petrology where rounded grains are common, hence the importance of the ellipsoid model.

For three shapes of rotational ellipsoid no corrections are necessary as the projected size is unique (Figure 3.34): (1) The projected size of a sphere equals its true size. (2) The projected length of oblate ellipsoids equals the major axis of the ellipsoid. (3) The projected width of prolate ellipsoids equals the minor axis of the ellipsoid. The length of prolate ellipsoids and the width of oblate ellipsoids are not unique and these measures should be avoided. For triaxial ellipsoids the outline length is strongly bimodal with peaks at the major and intermediate axes; the outline width is also bimodal with peaks at the intermediate and short axes. The complexity of these data necessitates more complex data reduction methods. A method similar to that of Saltikov (Section 3.3.4.4) can also be applied to these models.

The length of the outlines of both tablets and prisms are easy to interpret (Figure 3.35a): The outline length is close to the actual long dimension of the parallelepiped. The width of tablets, when normalised to the intermediate dimension, is widely dispersed and this measurement should be avoided

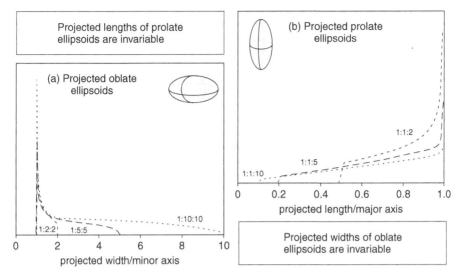

Figure 3.34   Projections of ellipsoids with random orientations (this study). Oblate ellipsoids are wide and short; prolate ellipsoids are tall and thin. Lengths of projections are divided by the major axis of the ellipsoid; widths are divided by the minor axis of the ellipsoid. Invariable lengths are always equal to the major axis of the ellipsoid and invariable widths are always equal to the minor axis of the ellipsoid. Invariable parameters are always the best choice as no stereological corrections are needed.

(Figure 3.35c). The normalised width of the prisms is simple and identical for the dimensions: it varies from 1 to 1.5 (Figure 3.35d). The normalised length of the prisms is widely dispersed and should be avoided (Figure 3.35b). Blocks with shapes between tablets and prisms have complex distributions and a method similar to that of Saltikov can be used.

3.3.6 Verification and correction of grain size distributions using the volumetric phase proportion

Higgins (2002a) has shown that it is easy to verify that CSDs have been calculated correctly by comparing the volume of a phase present in the rock calculated using stereologically exact methods (such as phase area proportion; see Section 2.7) with that calculated from the CSD.

The volumetric proportion of phase, V, is calculated by integration of the volume of all the crystals:

$$V = \sigma \int_0^\infty n_V(L)L^3 \mathrm{d}L$$

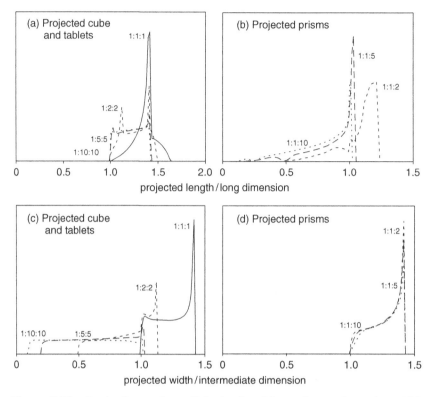

Figure 3.35　Projections of parallelepipeds with random orientations (this study). Projected width and length are the dimensions of the smallest rectangle that can be fitted around the projected outline.

where σ = shape factor = the ratio of the crystal volume to that of a cube of side L (see below). Size is defined here as the length of the longest axis of the smallest rectangular parallelepiped that encloses the crystal.

The shape factor, σ, is expanded into a more applicable form:

$$\sigma = [1 - \Omega(1 - \pi/6)]IS/L^2$$

where Ω is the roundness factor, which varies from 0 for rectangular parallelepipeds to 1 for a triaxial ellipsoid (Higgins, 2000). Obviously most crystals have a more complex shape, but we are concerned here just with a statistical measure.

For CSDs with discrete size intervals these equations can be modified to

$$V = \sigma \sum n_V(L_j)L_j^3 W_j$$

The summation is for all intervals j. Hence the volumes of a phase can always be calculated from a CSD. It should be remembered that the natural logarithm of the population density is plotted on a classical CSD diagram.

If CSDs are straight on a classical CSD diagram of ln(population density) versus size (see below; Marsh, 1988b) the equation can be integrated to

$$V = 6\sigma n_V(0)C^4$$

where $n_V(0)$ = intercept and C = characteristic length = $-1/$slope. It should be emphasised that this equation is for straight CSDs and that volumetric phase proportions calculated in this way are sensitive to an incorrect choice of the shape factor.

These equations are a useful way to verify that CSDs have been correctly converted from two-dimensional data, by comparing the measured value of the phase proportion with that defined by the CSD. However, caution must be exercised as the uncertainty in phase proportion estimated from CSDs can be significant. This is because much of the phase volume is contributed by the largest crystals. These are not numerous and hence their population density is not well known.

It is also possible to correct the CSD so that the volume calculated from the CSD, V_{CSD}, matches the measured phase volume, V_M. This process closely resembles normalising major element analyses to 100%, and similarly must be used with caution. We assume that error in V_{CSD} arises from a systematic error in the relationship between the crystal size used in the CSD and the crystal dimensions. For instance the crystals may be partly convex, hence the simple ellipsoid or parallelepiped models cannot be applied. These errors can be corrected if the true size of each crystal is multiplied by a correction factor F:

$$F = \sqrt[3]{V_M/V_{CSD}}$$

This correction is available in the program CSDCorrections (see Appendix).

3.3.7 Graphical display of grain size distributions

Grain size distributions can be expressed graphically in many different ways. Such diagrams are useful to show how well a grain size distribution corresponds to various theoretical size distributions. Some authors have confused the numerical grain size distribution with the graph used to display such distributions (e.g. Pan, 2001): obviously, the grain size distribution of a population does not change when different diagrams are used to display it. Various theoretical size distributions can be verified by graphical or numerical methods (see Section 2.5).

3.3.7.1 Size limits of measurements

All analytical methods are limited in the size of the grains that can be accurately measured (see Section 2). The upper size limit of all methods is the number of large grains that are in the largest size bin. The precision needed will vary with the goal, but for diagrams that use a logarithm of the size a minimum of 4 crystals is generally sufficient, but not ideal. In general, the measured maximum grain size can be increased by augmenting the volume or area measured.

The smallest grain that can be measured is equivalent to the limit of quantification in chemical analyses. The grain size distribution below this value is undefined. It may be necessary to combine different analytical methods to get an adequate range in measurable sizes (see Section 2). It is very important that size limits of the analytical technique are discussed in a study. If no crystals are recorded below a certain size, then it must be made clear if small crystals are absent or if the cut-off is an artefact of measurement.

3.3.7.2 Semi-logarithmic: 'CSD diagram'

In igneous petrology the natural logarithm of the population density of crystals is commonly plotted against linear crystal size, following the initial work of Marsh (1988b). This diagram has become known as the CSD diagram (Figure 3.36).

Population density $n'_V(L)$ is defined as the number of crystals per unit volume within a size interval ΔL as ΔL tends to zero (Marsh, 1988b). This is not very practical as the number of crystals is limited. Marsh (1988b) suggested that it is better to use the cumulative number of crystals greater than size L, $N_V(L)$. This is a stable, continuously increasing function from which the population density can be simply derived:

$$n'_V(L) = \frac{dN_V(L)}{dL}$$

This definition can be used if the true size of all crystals is known, which is the case for large 3-D datasets derived by tomography or serial sectioning. However, in many CSD studies data are gathered from two-dimensional sections and the data transformed to discrete intervals of three-dimensional sizes. We then need a definition of population density based on size intervals.

In a size interval j, the population density, $n'_V(L_j)$, is the number density of crystals in the size interval, $n_V(L_j)$, divided by width of the interval, W_j.

$$n'_V(L_j) = \frac{n_V(L_j)}{W_j}$$

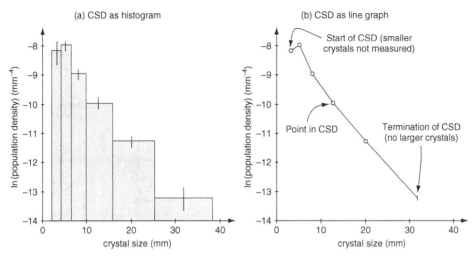

Figure 3.36 A simple semi-logarithmic 'CSD diagram' following Marsh (1988b). (a) The diagram is actually a histogram as data are integrated over the width of each size bin. Each size bin has an uncertainty associated with it indicated by the uncertainty bar. (b) CSDs are commonly drawn as line graphs, by connecting the centre of the top of each histogram bar. Higgins (2003) has suggested that the nature of the ends of the CSDs be clearly indicated: a vertical bar shows that there are no larger (or smaller) crystals. A circle just indicates a data point. A circle at the end of the CSD indicates that the termination is just an artefact of measurement as we have not examined smaller or larger crystals.

Note that the units of population density, $n'_V(L)$, are length^{-4}, as the volumetric number density of crystals, $n_V(L)$, has the units of length^{-3} and it is divided by a length, the interval width W_j.

A CSD is said to be continuous if it has no empty bins. For a continuous CSD the value of population density will change smoothly for different bin widths. If gaps (empty bins) appear owing to scarcity of data then either more data should be acquired or the bin widths should be widened until every bin has an adequate number of crystals in it. However, there is no intrinsic reason why all CSDs should be continuous and so we must be able to accommodate such gaps. Although commonly drawn as line graphs or as a series of unconnected points, CSD diagrams are actually histograms (Figure 3.37). Therefore, the line type CSDs should not be connected across empty bins. A suggestion for clarifying this problem is shown in Figure 3.37b.

If the CSD is linear on this diagram then it can be regressed and the intercept and slope determined. Some regressions use the uncertainty in each point as weighting and calculate the goodness-of-fit parameter Q (see Section 2.5.2; e.g. CSDCorrections 1.3).

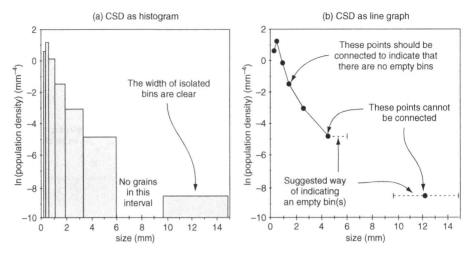

Figure 3.37   The problem of 'empty' bins in CSD diagrams. (a) If the data are presented as a histogram, then the width of the empty bin is clear. (b) On a conventional CSD diagram the width of the empty bin is not clear. It is suggested here that the widths of terminal (end) and isolated bins are indicated by a horizontal dashed line ending a vertical bar.

The slope (m) is sometimes transformed into the characteristic size, C, as follows:

$$C = -1/m$$

This parameter is easier to comprehend than the slope, as it has the units of length. Indeed, for a straight CSD that extends from zero to infinite size, the characteristic size is equal to the arithmetic mean size of all the crystals.

3.3.7.3 Bi-logarithmic: 'Fractal'

The mathematical concept of fractional dimensions – fractals – is recent, but has had a remarkable impact on the natural sciences, particularly in geology (Mandelbrot, 1982, Turcotte, 1992). Many natural features are thought to be self-similar: that is, the texture remains the same at all magnifications. This behaviour can be described by a fractal dimension.

Fractal (self-similar) distributions can be identified using a bi-logarithmic diagram – for instance log (number of grains larger than r) against $\log(r)$, where r is the size of the grains (Figure 3.38). A similar diagram of ln (population density) against ln (size) can also be used. It is very important to use a wide range of crystal sizes, otherwise such a fractal behaviour cannot be identified (Pickering *et al.*, 1995). The slope of the line can be used to establish the fractal dimension of the distribution. If the line has several well-defined slopes then the distribution may be described as multifractal.

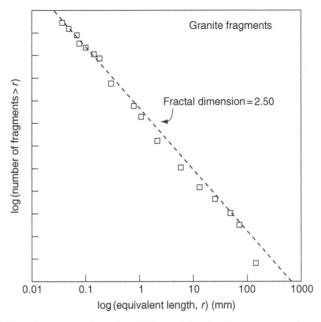

Figure 3.38 An example of a bi-logarithmic diagram used to examine possible fractal size behaviour. The size parameter (equivalent length, *r*) is the cube root of the particle volume. It is equivalent to the side of a cube of equivalent volume. The points on this diagram could be linked up as the distribution is continuous, but this has been omitted for clarity. After Turcotte (1992).

3.3.7.4 Lognormal and loghyperbolic

Lognormal distributions by mass or volume of grain sizes are commonly proposed in sedimentary petrology (Lewis & McConchie, 1994b). Cashman and Ferry (1988) proposed that metamorphic rocks also have a lognormal by number CSD. Such a lognormal by number model has also been proposed for CSDs of crystals grown at low temperatures in aqueous systems (Eberl *et al.*, 1998, Kile *et al.*, 2000). It should be mentioned that chemical abundances are commonly thought to follow a lognormal by number distribution, although recent analysis suggests that this is rarely true (Reimann & Filzmoser, 2000).

The parameter that has a normal or lognormal size distribution should always be specified. In sedimentology the parameter is generally mass, but volume may be used also. In other applications the parameter is the number of grains. A distribution that is lognormal by volume will not be lognormal by grain numbers (Jerram, 2001).

Some authors have used linear frequency diagrams to verify lognormal distribution (Eberl *et al.*, 1998); this is not the best solution, as deviations

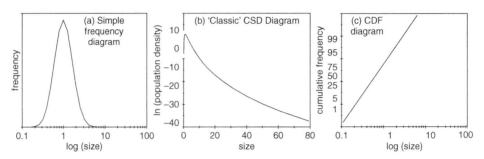

Figure 3.39   (a) Lognormal grain size distributions are not clearly displayed on a simple frequency diagram. (b) They are difficult to recognise on a semi-logarithmic 'CSD diagram'. (c) The clearest graph is of the cumulative distribution function (CDF) against log (size).

from the ideal distribution are not clearly shown, especially for sizes far from the mean (Figure 3.39). The best way to verify if data have a lognormal size distribution is to use a graph of cumulative distribution function (CDF) of volume, mass or number against log (size). Cumulative data can be transformed using the inverse function of the standard normal cumulative distribution available in spreadsheet programs (e.g. Excel function NORMSINV; see Appendix). Sedimentologists generally specify size in phi (Φ) units where

$$\Phi = -\log_2(L)$$

and *L* is in millimetres.

On this diagram a lognormal distribution plots as a straight line. Mixtures of lognormal distributions can sometimes be identified as two or more straight line segments. If intersection data are used then it is very important that conversion to size distribution is done using stereologically correct methods (see Section 3.3.2; e.g. Higgins, 2000) and that the data extend for several standard deviations on either side of the mean. Phi diagrams cannot be recalculated to population density diagrams unless the volume measured or the total phase volume is documented.

A number of sedimentologists have proposed that grain sizes follow a loghyperbolic distribution (Christiansen & Hartman, 1991). Normal distributions have the form of a parabola, when plotted as the log (probability density) against size; loghyperbolic distributions are hyperbolas in the same space. Whereas a normal or lognormal distribution is defined by two parameters, the mean and standard deviation, a loghyperbolic distribution needs four parameters. Very well defined size data are needed to verify if sizes follow a loghyperbolic distribution. Much sedimentological size data may not warrant this level of treatment as the methods used to obtain the data commonly have systematic biases. There is no particular graphical solution that is amenable to verify this distribution. Another distribution that has proved popular is the Weibull distribution (Murthy *et al.*, 2004). This function can have many very different forms and is popular with modelling of failure in mechanical systems. Again there are no particular graphs that can linearise this function.

3.3.7.5 Relative frequency and other methods

Many studies express size data as a simple histogram of relative or absolute number frequency or per cent volume or mass against size (Figure 3.40a). This class of diagrams has the benefit of simplicity, but few other advantages. The height of each bar in the histogram clearly depends on the width of the size interval: if the width is doubled then the number of grains in that interval will increase also. If size intervals are not constant this can create a false view of the data. It is also difficult to see variation in frequency for very small or large crystals. If this type of diagram must be used then population density (frequency / interval width) should replace frequency, so that the height of the histogram bars will not be affected by the width of the intervals (Figure 3.40). Data presented on these types of diagram cannot be converted to a classical CSD diagram without knowledge of the volume of the sample.

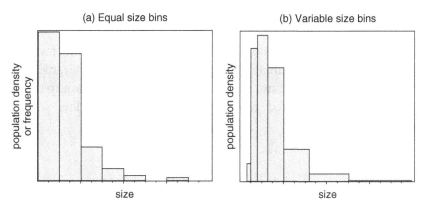

Figure 3.40   Plot of the same grain population on two simple, linear size distribution graphs. (a) Equal sized bins are used to express CSDs in many studies. For equal sized bins, frequency and population density graphs are identical. Here, the graph conceals much detail for small grains, and there are empty bins for large grains. (b) If the population density is used then the bin widths do not have to be constant and they can vary with the quantity of data. Here logarithmic-sized bins reveal aspects of the distribution concealed by fixed width bins.

3.3.8 Overall grain size parameters: moments of number density

Moments are parameters extracted from the size distribution that are used for describing overall aspects of the grain population. The material is assumed to be homogeneous and without preferred orientation (fabric). The moments of a number density distribution, M_0, M_1 etc., are defined by

$$M_i = \int\limits_0^L L^i n_V(L) \mathrm{d}L$$

M_0 = total number of grains per unit volume

M_1 = total length of grains per unit volume

Higher moments must accommodate a shape factor for departures of the grains from cubes.

$M_2 = C_A{}^*$ total projected area of grains per unit volume. The shape factor C_A is equal to the mean projected area of the grain divided by L^2. This moment has been misconstrued by some authors as being related to the total surface area of the grains. The total surface area can be obtained precisely by other methods without the need for a shape factor (see Section 2.4.2)

$M_3 = C_V{}^*$ total volume of grains per unit volume. The shape factor C_V is equal to the volume of the grain divided by L^3. This moment has been used to verify that CSDs have been calculated correctly (see Section 3.2.6). Again this parameter can be obtained precisely by other methods without the need for a shape factor (see Section 2.4.2).

If the size distribution function is known (variation of number with size) then the moments can be evaluated in terms of other textural parameters (e.g. Marsh, 1988b). Moments are not generally very useful, as the same parameters can commonly be evaluated more precisely with much less work (see Section 2.4.2).

3.4 Typical applications

The applications chosen here are representative of the diversity of materials and problems that have been studied so far and extend two previous reviews of the subject (Cashman, 1990, Marsh, 1998). However, number of studies on any rock type is very limited and hence much work is still exploratory. The scientific quality of publications in this field is extremely variable: some authors continue to use discredited conversion methods even though they know that they give incorrect results or have not understood the nature of the conversion problem. Other authors do not specify how they converted intersection data, and hence their results are of little use.

Many older CSDs were calculated using conversion equations, such as the 'Wager' or 'Kirkpatrick' method that are now known to be inaccurate. These data can be recalculated if the shape of the crystals, the nature of the rock fabric and the orientation of the section to the fabric can be estimated (see Section 3.3.4.5). If the raw data are available then this is simple, but more commonly the published graphs must be measured and back calculated (a few early CSDs have been recalculated in this way in Higgins, 2000). The greatest error in the 'Wager' method is generally in the intercept of the CSD and the least in the slope. In addition, the abundance of smaller crystals is much less accurate than that of large crystals. Where necessary, and possible, data presented here have been recalculated for the more important papers using stereologically correct methods (Higgins, 2000) and converted to millimetre units and natural logarithms.

Another possible pitfall in reviewing published data is the units used in CSDs, which are not always explicitly indicated. In some CSDs the population density is calculated in terms of cm^{-4}, even if the size axis is in millimetres. In this case:

$$\ln(\text{population density in } cm^{-4}) = \ln(\text{population density in } mm^{-4}) + 9.2$$

In addition some authors use a logarithmic base 10 scale for the population density, instead of natural logarithms (base e, $2.718\ldots$).

CSD data can be presented on many different types of diagram. If none is specified then it must be assumed that a 'classical' CSD diagram of ln (population density) against size was used.

In the available literature growth rates are expressed in a number of units that are not easy to comprehend. Here are a few useful conversions for a typical growth rate: $10^{-10}\,cm\,s^{-1} = 10^{-9}\,mm\,s^{-1} = 1\,pm\,s^{-1} = 0.031\,mm\,a^{-1}$. The conventional atomic radius of oxygen is about 120 pm, hence this corresponds to one layer of oxygen atoms every four minutes.

3.4.1 Mafic volcanic rocks

3.4.1.1 Plagioclase in a basaltic lava lake

The study by Cashman and Marsh of the Makaopuhi lava lake (Cashman & Marsh, 1988) was the first geological application of the CSD theory developed by Randolph and Larson (1971) and modified by Marsh (1988b), and is probably the most quoted CSD paper. The Makaopuhi lava lake was on the flanks of Kilauea Iki, Hawaii, and was filled in 1965 with basalt initially containing about 4% plagioclase, as well as olivine and augite. The lava lake was drilled repeatedly during its solidification; hence a time-sequence of textural development in the upper part was sampled. CSDs of plagioclase, ilmenite and magnetite were determined from six samples in a single drill hole that had total crystal contents from 20% to 80%. Cashman and Marsh (1988) measured the lengths of a large number of crystals, 3000–5000 per section, and reduced the data using the 'Wager' method (see Section 3.3.4.6). Plagioclase data have been recalculated in Figure 3.41.

The corrected CSDs are slightly more curved than the CSDs published by Cashman and Marsh (1988), but can still be considered almost straight (Figure 3.41). Overall, the new CSDs have shapes and relative differences similar to those of the original data, but the actual population densities are different. The sample with the lowest crystal content has the steepest slope and this decreases regularly with increasing crystal content. The intercept of the regression is constant. Cashman and Marsh (1988) applied the theory of

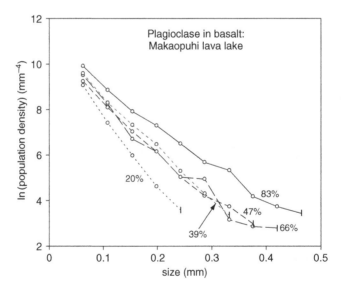

Figure 3.41   Plagioclase in basaltic magma from the Makaopuhi lava lake, Hawaii (Cashman & Marsh, 1988). Data were recalculated from graphs of the reduced data in Cashman and Marsh (1988) and Cashman (1986) using a crystal shape of 1:3:4, block shape and a massive texture. Some data intervals have been amalgamated to improve precision and fill empty bins. The intersection data have been cut off at 0.03 mm, as the crystals are dominantly in projection below this size and CSD is undetermined (see Section 3.3.2, Figure 3.25). With the new corrections the slope of the CSD was reduced by a factor of about 1.5 and the intercept increased by a factor of 15. Crystallisation is shown in per cent.

Marsh (1988b) to these results, although crystallisation in a lava lake is clearly not a steady-state process. Using the methods of Cashman and Marsh (1988), and the slope of the corrected CSD, the plagioclase growth rates were 3.5×10^{-10} to 6.5×10^{-10} mm s^{-1} and the nucleation rate was 3.6×10^{-5} to 7.2×10^{-5} mm^{-3} s^{-1}.

It is profitable to look at the changes in the CSD with solidification. Cashman and Marsh (1988) interpreted their data to indicate that both the nucleation and growth rates deceased with solidification. The data for growth rates are much more secure than those for nucleation rates, which depend on just one early sample. They considered that this was in response to linearly increasing undercooling, and that each sample developed under constant conditions. However, increasing undercooling tends to increase both growth rates and nucleation rates (e.g. Markov, 1995). It is perhaps easier to think of the rock textures seen here as developing sequentially and hence the samples represent a sequence of textural development during solidification (Higgins, 1998). Coarsening is an important process in petrology (see Section 3.2.4) and I have proposed that it may account for some aspects of the textures of these rocks (Higgins, 1998). However, closed system coarsening cannot account for the progressive solidification of the system and hence overall growth must also be important.

3.4.1.2 Plagioclase in basalt flows

Mt Etna is a large volcano that has been erupting porphyritic hawaiites for the last 200 years. Armienti *et al.* (1994) examined the CSDs of plagioclase, olivine, clinopyroxene and oxide minerals in a range of samples from the 1991–3 eruption. They measured crystals using optical and electronic methods and converted the data using a modified Saltikov technique. All minerals have strongly curved CSDs, with steep slopes for microlites (down to 30 µm) and shallow slopes for large crystals. However, linear size intervals were used in this study, which resulted in many empty bins for large crystals. If the bins are widened to eliminate this effect then the slope of the right side is much greater than presented. Armienti *et al.* (1994) proposed three intervals of crystallisation: the largest crystals nucleated and grew at depth in the magma chamber. As the magma rose in the conduit further nucleation and growth occurred. The microlites formed during quenching of the magma at the surface. However, mixing of two straight CSDs produces a curved CSD similar to that seen here (see Section 3.2.3.3), hence only two different periods of crystallisation are necessary. The similarity of the CSDs of this eruption to those of earlier eruptions was interpreted to mean that the main features of the volcanic feeder system had not changed recently.

The a'a and pahoehoe parts of lava flows have both distinctive surfaces and internal rock textures. Textural work so far on this problem has been somewhat ambiguous. Sato (1995) examined the CSD of two samples, one of each texture. Unfortunately the samples were from different eruptions. Nevertheless, he found important differences in their CSDs. The a'a sample has a steep CSD that did not extend to large sizes, whereas in the pahoehoe sample the CSD was much more shallow. He attributed this difference to the initial temperature of the magmas, before degassing and eruption, with the pahoehoe having a higher initial temperature. However, this does not seem to fit with the commonly observed transition from pahoehoe to a'a along some lava flows. Polacci *et al.* (1999) only looked at crystal number densities instead of CSDs. They concluded that crystallisation was occurring within the lava flow itself. Cashman *et al.* (1999) only examined the total crystal number by area and concluded that the magmas were well stirred and hence there were no delays to nucleation. Clearly there is room for more comprehensive studies.

Sections of many thicker basalt flows reveal two parts that are clearly defined by the joint patterns: at the base is the colonnade, with regular, commonly vertical, columnar joints; above it lies the entablature with more closely spaced, complex joints. Jerram *et al.* (2003) examined the CSDs of the two components in the Holyoke basalts (Figure 3.42). Both CSDs are straight,

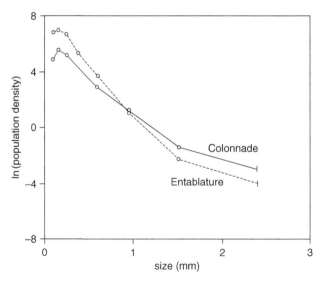

Figure 3.42   Holyoke basalt flow (from Jerram *et al.*, 2003). The entablature is the upper irregular part of the flow and the colonnade is the part with parallel columnar joints. The kink in the CSDs probably represents a change in the crystallisation conditions.

except for large sizes (phenocrysts), suggesting a two-stage crystallisation process. The colonnade has a shallower slope, which can be interpreted as a longer residence time, if a steady-state model is assumed. Jerram *et al.* (2003) calculated residence times of 8 and 11 years for the entablature and colonnade respectively for a growth rate of $5 \times 10^{-10}\,\mathrm{mm\,s^{-1}}$. This gives reasonable solidification times for the whole flow. The difference between the CSDs of the entablature and colonnade may also be interpreted in terms of coarsening (Higgins, 1998): The entablature then represents an earlier texture than the colonnade, which had more time to coarsen.

3.4.1.3 *Plagioclase growth rates in basaltic magmas.*

Cashman (1993) reviewed plagioclase growth rates in basaltic systems. Her study was based on well-known volcanic systems in which the residence time can be estimated from the history of the volcano or other parameters and dykes in which cooling parameters can be easily calculated. For samples with CSDs she used the slope of straight CSDs to get values for the growth rate. In other samples she relied just on the mean crystal size and total number density. She also examined experimental results and concluded that nucleation in natural systems was generally heterogeneous.

Her main conclusion was that growth rates varied with cooling rate, but that the parameter was not very sensitive (Figure 3.43). A cooling time of 3 years

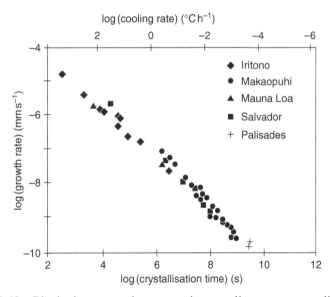

Figure 3.43 Plagioclase growth rate against cooling rate or cooling time (Cashman, 1993).

($=10^8$ s $=$ cooling rate of $0.008\,^\circ\mathrm{C}\,h^{-1}$) gives a growth rate of 10^{-9} mm s^{-1} and a cooling time of 300 years ($=10^{10}$ s $=$ cooling rate of $0.00008\,^\circ\mathrm{C}\,h^{-1}$) gives a growth rate of 10^{-10} mm s^{-1}. Although some of the CSDs used by Cashman were measured from sections and not calculated with the appropriate equations, the errors in the slope are small compared to the range in the slope values. Such growth rates should be only used in a general way to determine the residence time or other parameters to within an order of magnitude.

It is not clear how these growth rates can be modified for more silica-rich magmas. In general such magmas are more viscous and hence have lower diffusion rates. However, this is critically dependent on the water content of the magmas at depth, which is generally poorly known. Cashman (1992) estimated that the crystallisation rate for plagioclase in dacite was 5–10 times less than in basalts. Higgins and Roberge (2003) have proposed that crystallisation is not continuous but cyclic in an andesite magma chamber. If this is generally true then it may not be possible to determine crystal growth rates from natural systems.

3.4.1.4 Plagioclase in andesite

Mt Taranaki is an active andesite stratovolcano in New Zealand (Higgins, 1996a). Lavas on the flanks of the volcano contain abundant, generally euhedral plagioclase crystals, comprising up to 30% of the rock on vesicle-free basis. Many crystals have concentric zones of dusty inclusions, which are commonly thought to represent periods of partial melting or re-heating. The CSDs define a broad curved band, but each individual CSD is much straighter than the overall trend suggests. Indeed, individual CSDs are almost linear down to small crystal sizes, suggesting that nucleation and growth dominated (Figure 3.44a). If the crystals grow in a steady-state magma chamber with a growth rate of 10^{-10} mm s^{-1} then characteristic lengths indicate that the earliest lavas (Figure 3.44b; Staircase group) had a residence time of $\sim$50 years. Subsequent lavas in the Castle and Summit groups had slightly longer residence times of 50–75 years. The youngest magmas, from both Egmont Summit and the Fantham's peak, a secondary vent on the side of the volcano, have the shortest residence times of $\sim$30 years. Variations in residence time may reflect changes in the magma chamber shape or depth, or the temperature of the surrounding rocks.

The Dome Mountain lavas comprise a comagmatic pile of some twenty flows with a total thickness of about 300 m (Resmini & Marsh, 1995). Most flows are andesitic, but there is some variation to more mafic and felsic compositions. Most of the plagioclase CSDs show no sign of crystal fractionation or accumulation, hence they can be used to give magma storage times. If a growth rate of 10^{-9} mm s^{-1} is assumed then most magmas have residence

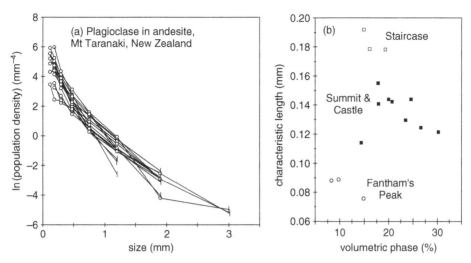

Figure 3.44 Plagioclase in andesite lavas from Mt Taranaki (Egmont Volcano), New Zealand. The original data from Higgins (1996a) were recalculated and published in Higgins (2000). (a) In a conventional CSD diagram it is not easy to discern variation trends between samples. (b) Although the CSDs are slightly curved it is still profitable to calculate the characteristic length ($-1/$slope) and plot this against the dense-rock volumetric phase per cent of plagioclase, calculated from the total area of the crystal intersections and corrected for vesicles. Three groups of lavas clearly stand out on this diagram.

times of 1.5 to 4 years, except for two of the lower lavas with residence times of 4 to 9 years. The residence times correlate inversely with nucleation density and broadly correlate inversely with silica content: high-residence-time, low-silica lavas occur at the base of the pile, whereas low-residence-time, high-silica lavas occur at the top of the pile. One interpretation of the residence time data is that the lavas are from an open system of more or less constant residence time but varying spatially in system locations; e.g. a magma chamber in which steady-state recharge and eruption may have been reached. Alternatively, these lavas may represent small samples, closely spaced in time, of a slowly cooling, large batch or reservoir of magma. The greater residence times associated with the lower flows may indicate the increased amount of time necessary to form the conduit of the magma-volcano system by the oldest magma or simply that part of the system with the highest liquidus temperature.

The Soufriere Hills volcano, Montserrat started to erupt at the end of 1995 after a hiatus of 300 years and continues to this day. The volcano is dominated by a dome of andesite that sheds pyroclastic flows as it inflates. Typical dome fragments are glassy and rich in plagioclase crystals – up to 35% for crystals larger than 0.03 mm. The pyroclastic material is texturally similar, but vesicular. A particular aspect of these rocks is the abundance of large amphibole oikocrysts, up to 12 mm long, many of which are loaded with plagioclase crystals. Oikocrysts seal in early textures and preserve aspects of them for later examination. Hence, it is possible to see a sequence of textures in such rocks that can show how the CSD has varied with time (Higgins, 1998). Plagioclase CSDs in both dome fragments, pyroclastic rocks and within amphiboles were examined by Higgins and Roberge (2003).

The texture of the material outside the oikocrysts, the matrix, largely preserves the last state of the magma before eruption whereas the oikocrysts preserve earlier textures. All matrix CSDs are strongly curved and broadly similar (Figure 3.45). CSDs of plagioclase from the oikocrysts are coincident with the CSDs of the matrix, but do not extend to the same sizes. In fact, there appears to be an almost continuous range of maximum crystal sizes in different matrix samples and oikocrysts. The instability of amphibole at depths above 5 km indicates that plagioclase crystallisation began below that level.

The oikocryst and matrix plagioclase textures cannot be related by simple growth of plagioclase. Plagioclase crystals must start small and numerous and progress towards small numbers of larger crystals. The overall process must

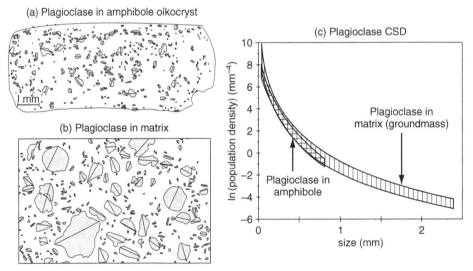

Figure 3.45 Plagioclase in andesites of the Soufriere Hills volcano, Montserrat (Higgins & Roberge, 2003). (a) Plagioclase chadocrysts in the amphibole give an early view of the texture in the magma chamber. (b) Plagioclase in the matrix gives a view just before the eruption – same scale as the oikocryst. (c) Plagioclase CSDs in both environments are summarised by envelopes that represent the range of a number of samples. The two groups of CSDs are similar for small sizes, but that of the matrix continues the curve to larger crystals. A similar variation can be seen in the individual matrix CSDs.

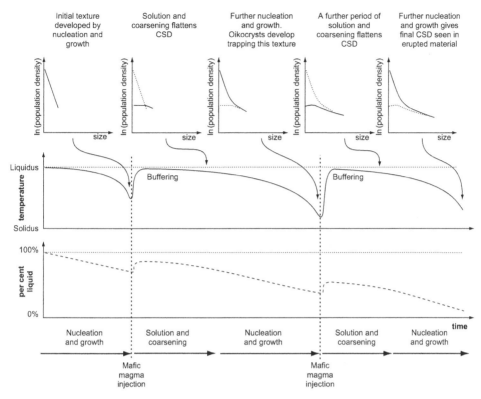

Figure 3.46   Soufriere Hills volcano, Montserrat. The solidification model of Higgins and Roberge (2003) involves repeated cycles of nucleation and growth, followed by coarsening.

progressively extend the CSD towards larger crystals. Such an overall process must involve both crystallisation and solution. Higgins and Roberge (2003) proposed a model that involves repeated cycles of crystal growth and coarsening (Ostwald ripening) in response to varying degrees of undercooling and superheat (Figure 3.46). The first cycle started when plagioclase crystallised in response to increasing undercooling in the magma chamber. This phase ended when the undercooling decreased. This could have been caused by addition of heat from mafic magma intruded into the magma chamber or by changes in pressure (depth) or volatile fugacity. Under these conditions coarsening began. Small crystals dissolved to feed larger crystals. Numerous repetitions of this cycle will yield the observed complexly zoned large crystals. At Montserrat the overall process has increased the total crystal content of the magma, but this may not necessarily happen elsewhere. Finally, small crystals grow in response to uplift that just precedes the eruption. Amphiboles may have also grown by the same process.

3.4.1.5 Olivine in picrite

The 1959 eruption of Kilauea Iki comprised a series of lava flows and created a lava lake that has only recently completely solidified. The magma was picritic, a very unusual composition for Hawaii. Mangan (1990) studied the CSDs of olivine in samples of the pyroclastic deposits of this eruption. The material is not very easy to measure because some of the olivine crystals are xenocrystic, and were derived from disaggregated mantle peridotites. Such crystals have solid-state deformation features which are lacking in the magmatic olivine. However, this distinction becomes more difficult for smaller crystals. The magmatic olivine crystallised in a magma chamber beneath the volcano. Mangan (1990) measured a small number of crystals from many different samples, but data have been summed here to improve the precision (Figure 3.47). The turn-down for small crystals may be an artefact of measurement. Although the CSD of magmatic crystals is slightly curved, a

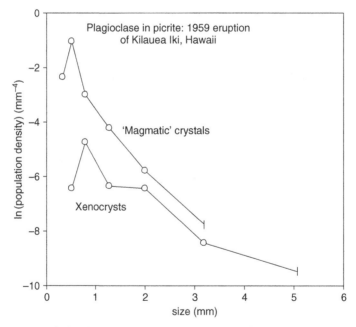

Figure 3.47   Olivine in picrite scoria from the 1959 eruption of Kilauea Iki (Mangan, 1990). Data have been recalculated from the original length measurements supplied by Dr Mangan using the shape of crystals in samples from the lava lake that developed later in the eruption that we have measured in our laboratory. For both populations of crystals the aspect ratios were 1:1.25:1.25, with a roundness of 0.5 and crystals were randomly orientated (massive fabric). 'Xenocrysts' show evidence of solid-state deformation; 'Magmatic' crystals lack such deformation.

characteristic length of 0.34 mm can be calculated. Mangan (1990) chose somewhat arbitrarily a growth rate for olivine of $10^{-9}\,\mathrm{mm\,s^{-1}}$ as this was what was used for olivine in the Makaopuhi lava lake (Marsh, 1988a). If we apply the Marsh (1988b) steady-state model, then this gives a residence time of 11 years. The significance of the CSD of the olivine xenocrysts is not clear and more data are needed.

3.4.1.6 Anorthoclase in phonolite

Dunbar *et al.* (1994) have looked at anorthoclase phenocrysts in phonolite from Mt Erebus, Antarctica. These crystals are up to 90 mm long and are very abundant. They sampled the crystals in volcanic bombs thrown out by the volcano. These bombs can be disintegrated easily and the crystals separated and measured using a ruler. The three samples measured all had straight CSDs down to 10 mm, the measurement limit (Figure 3.48). They did not attempt to extend this CSD to smaller sizes with a thin section; hence it is not clear if these rocks lack very small crystals.

Dunbar *et al.* (1994) concluded that anorthoclase had low nucleation rates but that it crystallised continuously through time. Anorthoclase should float in a phonolite magma, but the authors considered that the uniformity of the three CSDs did not support this idea. However, it seems to me there is no reason why flotation cannot produce uniform CSDs. The low overall crystal

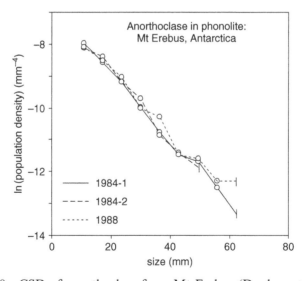

Figure 3.48   CSD of anorthoclase from Mt Erebus (Dunbar *et al.*, 1994). The crystals were extracted whole by disintegration of three volcanic bombs erupted in 1984 and 1988. The original data have been converted so that size equals length in mm using a width/length ratio of 0.33.

abundance and large sizes suggest that coarsening was important. Initially there may have been a high nucleation rate and many small crystals, but these were removed during coarsening. A key test would be to look at the abundance of small crystals – their absence would suggest coarsening. This could well be combined with accumulation by flotation.

3.4.2 Mafic plutonic rocks

Quantitative textural studies of plutonic rocks really started with Jackson's work on the Stillwater complex in 1961 (Jackson, 1961). He measured crystal size distributions of olivine, orthopyroxene and chromite, but did not record the area that he measured; hence his data cannot be completely compared with modern data.

3.4.2.1 Gabbro

The Oman ophiolite is one of the best exposed and preserved sections of the oceanic crust, and is thought to have formed at a fast or medium spreading ridge. Garrido *et al.* (2001) studied plagioclase in the lower and middle sections, whereas Coogan *et al.* (2002) examined plagioclase in the uppermost gabbros, just below the sheeted dykes. In both studies plagioclase CSDs were concave down, with straight right limbs and a clear lack of small crystals. The simple shape suggests a simple history: Coogan *et al.* (2002) proposed that a high-level magma chamber was filled with primitive, aphyric magma and that all crystals grew there in place. Garrido *et al.* (2001) considered that the lack of small crystals reflected coarsening during the early phases of the evolution of the textures: late reheating to 500 °C did not cause appreciable grain growth. Other authors have suggested that grain-boundary migration during coarsening can be halted by the presence of other phases (grain-boundary pinning). However, here it is shown that this process did not significantly limit the coarsening process. Garrido *et al.* (2001) divided up their gabbro section into two parts on the basis of the CSDs and interpreted this in terms of near-vertical isotherms close to the ridge axis.

The Kiglapait intrusion is one of the younger components of the Nain Plutonic Suite. It has a conical form about 30 km in diameter, and is about 8.5 km thick in the centre (Morse, 1969). It has been well studied because it is essentially undeformed and unaltered. The lower zone is composed of plagioclase and olivine; clinopyroxene joins the assemblage towards the top and marks the base of the upper zone.

Higgins (2002b) determined the CSDs of plagioclase, olivine ± clinopyroxene in a series of 13 sections from the base to the top of the intrusion

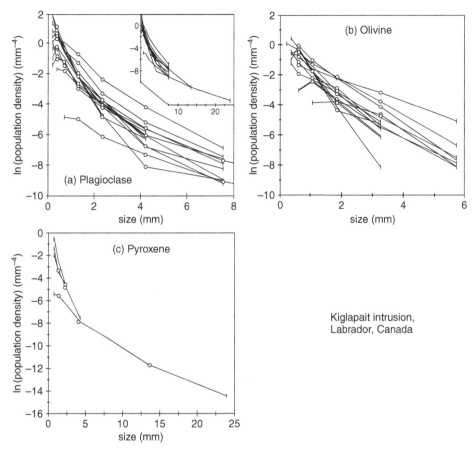

Figure 3.49 CSDs of plagioclase, olivine and clinopyroxene from the Kiglapait intrusion, Canada (Higgins, 2002b).

(Figure 3.49). All CSDs lack small crystals and are slightly curved to the right, concave up. However, the curvature is not as strong as first appears: the CSDs define a fan, as expected if there is little variation in mineral abundance (Higgins, 2002a). Most of the textural variation is the result of variable degrees of coarsening, following the Communicating Neighbours model (DeHoff, 1991). The degree of coarsening is lowest at the base, when the magma was first introduced into cool host rocks. Coarsening reaches a maximum when the intrusion was 20% solidified, and then decreases towards the top of the intrusion. This reflects the lower volume and hence heat production of the crystallising magma, which could be therefore dissipated more readily. There does not seem to be a correlation between the degree of coarsening of the olivine and plagioclase, suggesting that they were not always on the liquidus at the same time.

Part of the lower zone is well layered, in some cases with well developed mineral grading. If such layers were produced by crystal settling then we would expect to see grading in both the amount of the phases and their size distributions. One of these layers was examined in detail. A settling origin is strongly suggested by the sharp base rich in olivine, which grades up into a top rich in plagioclase. However, plagioclase and olivine CSDs from the base and top do not show the expected effect. This suggests that subsequent coarsening has obscured the grain size effects of crystal setting.

3.4.2.2 Troctolite and anorthosite

Higgins (1998) showed that the textures of plagioclase crystals within large olivine oikocrysts preserve a textural sequence of the formation of anorthosite. Such textures have been quantified in a troctolite/anorthosite from the Proterozoic Lac-Saint-Jean anorthosite suite, Canada (Figure 3.50). Plagioclase CSDs indicate that it initially nucleated and grew in an environment of linearly increasing undercooling, producing a straight-line CSD. During this phase, latent heat of crystallisation was largely removed by circulation of magma through the porous crystal mush. By about 25% solidification the crystallinity was such that it reduced, but did not eliminate, the circulation of magma, resulting in the retention of more latent heat within the crystal pile. The temperature increased until it was buffered by the solution of plagioclase close to the liquidus temperature of plagioclase. Nucleation of plagioclase was inhibited and conditions were suitable for coarsening of both plagioclase and olivine to occur.

Higgins (1998) considered that the shapes of the plagioclase CSDs fit better the Communicating Neighbours equation of coarsening, rather than the classic Lifshitz-Slyozov-Wagner equation (see Section 3.2.4), as do other examples drawn from the literature. If olivine started to nucleate at a higher temperature than plagioclase, then during the coarsening phase olivine would have been more undercooled than plagioclase, and would have had a higher maximum growth rate. Under these conditions olivine would coarsen more rapidly than plagioclase and engulf it. Hence the order of crystallisation determined from the textures would be the reverse of the order of first nucleation of the two phases, from equilibrium phase diagrams. Maintenance of the temperature near the plagioclase liquidus may also inhibit the nucleation and growth of other phases.

The upper part of the unmetamorphosed Sept Iles intrusive suite contains a series of anorthosites about 1 km thick (Higgins, 1991, Higgins, 2005). Parts of the anorthosite are very well foliated with abrupt transitions to massive anorthosite. In an early study I examined quantitatively the textures of one

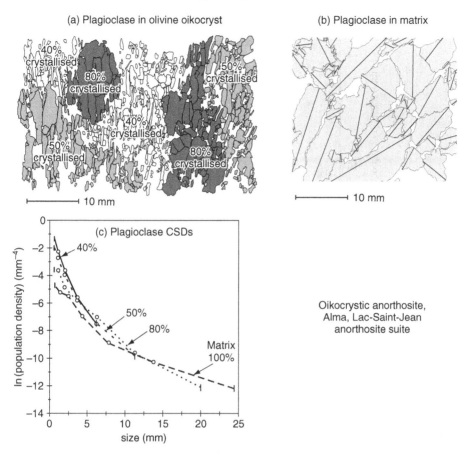

(a) Plagioclase in olivine oikocryst

(b) Plagioclase in matrix

(c) Plagioclase CSDs

Oikocrystic anorthosite,
Alma, Lac-Saint-Jean
anorthosite suite

Figure 3.50   Plagioclase in oikocrystic anorthosite (troctolite) from the Lac-Saint-Jean anorthositic suite (Higgins, 1998). (a) A large olivine oikocryst was subdivided into regions of varying plagioclase content: 40, 50 and 80%. They are considered to preserve a textural development sequence. (b) The anorthosite from between the oikocrysts is termed the matrix and is 100% plagioclase. (c) Plagioclase CSDs from the oikocryst regions and the matrix. Data were recalculated from the original data using a shape of 1:3:3 for the oikocryst samples and 1:1:1 for the matrix sample.

sample from each facies with the goal of determining the origin of the lamination (Figure 3.51; Higgins, 1991). One aspect that is particularly striking is the strong difference in crystal shape between the thin tablets of the laminated anorthosite (1:5:7) and the stubbier crystals of the massive anorthosite (1:2.5:4). The recalculated CSDs for the laminated and massive anorthosite are essentially collinear, but the massive anorthosite extends to larger sizes. There are no crystals smaller than 1 mm for the massive anorthosite and 2 mm for the laminated anorthosite. In the figure the CSDs terminate at 2–3 mm

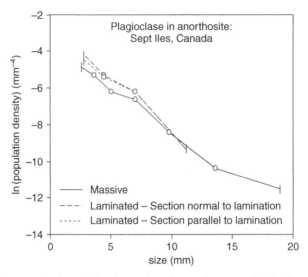

Figure 3.51   Plagioclase in laminated and massive anorthosite from the Sept Iles intrusive suite (Higgins, 1991). The original data have been recalculated using the methods of Higgins (2000) and a shape of 1:2.5:4 for the massive sample and 1:5:7 for the laminated sample. Two orthogonal sections of the laminated sample were analysed: It is gratifying that they give very similar CSDs as there is, of course, only one actual CSD for a phase in a rock sample.

because of analytical difficulties. The similarity of the CSDs indicates that the anorthosite of the two facies crystallised in the same pressure, temperature and time environment, but that other processes were responsible for producing the foliation and differences in crystal shape. The most likely candidate is simple shear which can produce the foliation and also remove magma around the growing crystal depleted in the plagioclase component (see Section 5.2.2.2).

3.4.2.3 Impact melt rocks

The Sudbury igneous complex is the melt sheet produced by a large meteorite impact. The overall composition is a norite, but the body is differentiated. It may have been superheated when first produced. Zieg and Marsh (2002) have reported CSDs from this fascinating unit, but unfortunately give few details. They state that the plagioclase CSDs are linear, but do not present any actual CSDs. It is also not clear if small crystals are absent, as in many plutonic rocks. They have shown however, that the characteristic length of the crystals increases systematically with height in the melt sheet by a factor of two. They used these data and a simple cooling model to calculate an effective growth rate of 1.5×10^{-13} mm s^{-1} and a cooling time of 1.8×10^{12} s (57 ka). These values are approximately collinear with the data obtained from volcanic rocks by Cashman (Figure 3.43; 1993).

3.4.3 Silicic volcanic rocks

3.4.3.1 Plagioclase in dacite

Phenocrysts and microlites of plagioclase from the 1980–6 eruptions of Mt St Helens, USA, were examined by Cashman (1988, 1992). Both phenocrysts and microlites have straight CSDs and appear to have similar slopes, although this is difficult to verify as different measurement techniques were used in the two studies. The CSDs of microlites were combined with geophysical estimates of the residence time to calculate growth and nucleation rates. Cashman considered that the growth rate only varied by a small amount for different values of undercooling, but that the nucleation rate was strongly affected. This confirms standard nucleation models. She proposed a two-stage crystallisation model in which most of the phenocrysts nucleated at depth. At the end of the initial part of the eruption the magma resided in a chamber at a depth of 4–5 km. Further growth in this chamber was caused by coarsening, rather than nucleation of new crystals.

The Kameni volcano is part of the Thera (Santorini) volcanic complex, an active volcano in the eastern Mediterranean (Higgins, 1996b, Druitt *et al.*, 1999). The dacite lavas of the Kameni islands have been erupted over the last 2000 years and form an unusually useful field area: they are fresh, easily accessible and well-dated. All samples are porphyritic: the most common megacryst phase is plagioclase (typical mode 12%), with clinopyroxene, ortho-pyroxene, magnetite, olivine and apatite, all set in a fine-grained to glassy matrix with plagioclase microlites.

None of the plagioclase CSDs are linear, but curved concave upwards (Figure 3.52: Higgins, 1996b). Hence, a simple model based on a single population of crystals, such as that of Marsh (1988b), cannot be applied. Higgins (1996b) proposed that such CSDs could be produced by mixing of two populations of crystals, each with overlapping linear CSDs, termed micro-lites and megacrysts. The two populations can then be examined separately. Application of the Marsh model (1988b) with growth rates of 10^{-10} mm s^{-1} gives residence times of 6–13 years for the microlites and 24–96 years for the megacrysts. These residence times broadly accord with those determined from diffusion rates in plagioclase (Zellmer *et al.*, 2000). If the residence times of the megacryst populations are projected back from the eruption times, then some of the times of emplacement of the megacryst magmas are found to coincide

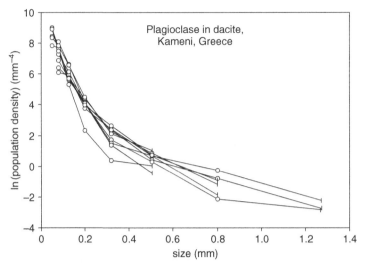

Figure 3.52 Plagioclase in dacites from Kameni volcano, Greece (Higgins, 1996b). Data have been reconverted using a massive block crystal shape with an aspect ratio of 1:4:4. These concave up CSDs can be modelled by a mixture of two populations of crystals, each with a straight CSD.

with the dates of earlier eruptions. This suggests that the megacrysts are 'leftovers' of earlier injections of new magma into a shallow chamber: that is some magma remains after each eruption and continues to crystallise. Each eruption is initiated by the emplacement of new magma with few or no crystals, which mixes with the older magma. Eruption followed 6–13 years after mixing. Such a model would suggest that some porphyritic magmas are products of a shallow magma chamber that is never completely emptied, just topped up from time to time.

Turner *et al.* (2003) examined basaltic andesites using CSDs and uranium disequilibrium and found a profound disagreement in the residence times determined by these two methods. They suggest that this is caused by incorporation of existing cumulates into new magmas, similar to the processes proposed for the Kameni dacites.

3.4.3.2 Microlites

Microlites occur in many volcanic rocks and their growth may trigger eruptions. For stereological reasons prismatic microlites are not easy to measure in section although they can be measured easily in projection. Castro *et al.* (2003) used optical scanning to measure the CSDs of pyroxene microlite prisms in obsidian from Inyo Domes, California. On a classical CSD diagram the CSDs are straight from 5 to 30 μm, with a turn-down for smaller sizes. This is not an artefact, as the lower size limit is 0.5 μm. On a simple frequency diagram the distributions appear to be lognormal. Such distributions were interpreted to reflect a short nucleation event, followed by growth.

Hammer *et al.* (1999) examined plagioclase microlites in early, pre-climatic dacite tephras from the 1991 Mt Pinatubo eruption. This material is particularly difficult to measure as the microlites are small and have shapes that vary strongly with size: the smaller crystals are tabular, whereas the larger crystals are prismatic. Nevertheless CSDs from 13 different surge-producing blasts were all straight. The slope was used to characterise the degree of plagioclase supersaturation: it was found to vary with the repose time between the eruptions (40 to >170 minutes). Overall crystallisation passed from a nucleation-dominated regime, with tabular crystals and steep CSDs to a growth-dominated regime with shallow CSDs and prismatic crystals.

3.4.3.3 Zircon in rhyolite tuffs and lavas

The CSDs of minerals in explosively erupted rocks are of great interest, as they can tell us a lot about the state of the magma chamber just before the eruption. However, crystals are commonly broken by the force of the eruption; hence there have been few studies to date. One way around this problem is to look at zircon and quartz crystals, which are generally robust enough to survive intact.

Bindeman (2003) liberated zircon and quartz crystals from the Bishop Tuff and other silicic ash-flow tuffs by crushing and acid digestion. The crystals were then measured under a microscope. This technique avoids problems introduced during stereological conversions and enables the measurement of a large number of crystals with a very wide range in size. Both quartz and zircon CSDs are concave down on a classical CSD diagram, with a straight right-hand side (Figure 3.53). However, the distributions are lognormal when plotted on a CDF diagram, and this is confirmed by other statistical tests. Bindeman (2003) interpreted these distributions using the single-stage size-dependent growth model of Eberl *et al.* (2002). However, as Bindeman (2003) points out, both zircon and quartz have complex zoning patterns, indicating periods of growth and solution. It seems more likely that the lognormal distribution is a result of the interplay of these processes rather than the simple model of Eberl *et al.* (2002), which in any case is not supported by theory. The concave down shape of the CSDs is found in other coarsened (ripened) rocks; hence coarsening may well dominate here, especially just before eruption.

CSDs of zircon in rhyolite lavas closely resemble those from tuffs described above: they are concave down and appear to have a lognormal distribution

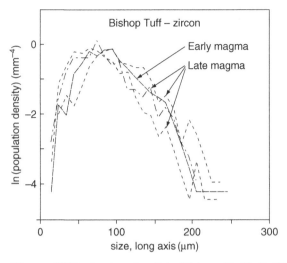

Figure 3.53   Zircon CSDs in the rhyolitic Bishop Tuff, California (from Bindeman, 2003). One sample from the early part of the eruption (top of the magma chamber) did not differ significantly from three samples from the later part of the eruption (base of the magma chamber).

(Bindeman & Valley, 2001). Bindeman and Valley (2001) used the slope of the right-hand side of the CSD to calculate a residence time of 100–200 ka for these zircons; however, as they point out, the steady-state CSD model may not be applicable here, the growth rate that they used is not well established and the zircons may be xenocrysts. Instead they suggest that coarsening may be important.

3.4.4 Silicic plutonic rocks

3.4.4.1 Orthoclase, plagioclase and quartz in granite porphyry

Mock *et al.* (2003) have done the first comprehensive quantitative textural study (size, shape, orientation and position) of a shallow level porphyritic pluton, using a vertical drill core that intersects the entire pluton. They examined the CSDs of orthoclase, quartz and plagioclase in slabs and found that all these minerals have remarkably straight CSDs for crystals larger than 1 mm. The abundance of crystals smaller than this limit is not clear. There is relatively little variation between samples within the laccolith. Mock *et al.* (2003) considered that these CSDs showed that these three minerals nucleated and grew continuously during magma ascent and emplacement; *in situ* crystallisation was not detected as the laccolith cooled fast after emplacement at such high levels. Minor changes in the textural parameters suggested that these were two episodes of magma injection into the laccolith.

3.4.4.2 K-feldspar in granodiorite

Euhedral potassium-feldspar megacrysts up to 20 cm long are a striking and relatively common component of many granitoid plutons, but their origin is still controversial (see the review by Vernon, 1986). Many geologists hold that the megacrysts are igneous in origin, that is, they grew from a granitic magma and are phenocrysts. However, a minority view is that megacrysts are post-magmatic and grew from a circulating water-dominated fluid. Megacrysts are fundamentally a textural phenomenon; hence lend themselves to CSD studies. CSDs of megacrysts in the Cathedral Peak granodiorite (CPG) were studied by Higgins (Figure 3.54a; 1999).

The CPG is the most voluminous part of the concentrically zoned Tuolumne Intrusive Suite, in the Sierra Nevada of California (Bateman & Chappell, 1979). This suite started with the emplacement of dioritic magma. Each of the following three phases was more felsic and was emplaced into the still-mushy core of the intrusion. An important component of the Cathedral Peak granodiorite, adjacent parts of the Half-Dome granodiorite and the Johnson Granite Porphyry is microcline megacrysts. The megacrysts are distributed

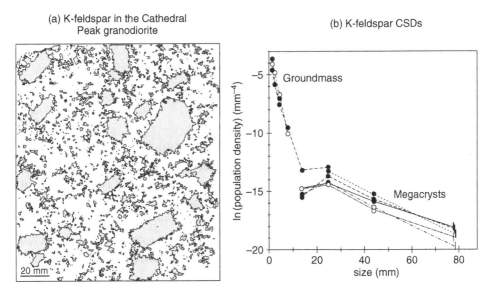

Figure 3.54 (a) K-feldspar in a slab from the Cathedral Peak granodiorite. The sample was stained, scanned and filtered for K-feldspar. (b) CSDs of microcline crystals (Higgins, 1999). Megacrysts from five outcrops were measured in the field with a ruler. Dimensions of groundmass microcline crystals in the slabs were determined by automatic image analysis. Open symbols are for areas with diffuse megacrysts whereas solid symbols are for sites with 'nests' of megacrysts.

heterogeneously with 'nests' and elongated areas up to a metre wide. The linear zones meander for tens of metres across the outcrop and there does not appear to be any contrast between vertical and horizontal sections. In general, three-dimensional sheets have linear intersections with surfaces whereas string-like forms have point intersections. Therefore, the surface distributions of the megacrysts suggest that the volumes rich in megacrysts have sheet-and string-like forms in three dimensions.

Somewhat surprisingly the CSDs of both the 'nests' of megacrysts and the diffuse areas have similar shapes, but the nests have higher overall population densities for all crystal sizes than the diffuse areas (Figure 3.54b). All CSDs have a peak at 25 mm. The smallest megacryst that could be reliably measured was 10 mm; hence the turn-down of the population density to the left of the peak is not an artefact. If the CSDs were straight down to 10 mm then the population density at 10 mm would have been 10 to 50 times higher, which is readily distinguishable from the actual data. The limb to the right of the peak is generally straight. The CSDs of the groundmass microcline are quite different from those of the megacrysts. All three CSDs are quite straight, right down to the smallest crystal that could be measured (1 mm) and much steeper than those of the megacrysts. One slab was sufficiently large that the abundance of smaller megacrysts could be determined, although the small numbers of large crystals did not give a high precision (Figure 3.54a). The compatibility of the two methods is confirmed by the overlap of the CSDs.

3.4.5 Other undeformed minerals

3.4.5.1 Diamonds

The value of diamonds in a deposit is critically dependent on the distributions of sizes and quality of the crystals. Although the size distributions of diamonds are commonly measured this information is usually secret. Of course, it is not always clear if diamonds are phenocrysts or xenocrysts. However, their CSD should tell us something about their regions of origin, the mantle. In this environment one would expect that diamonds are coarsened, unless protected inside large silicate grains. Hence, the relationship between microdiamonds and macrodiamonds should be very interesting.

Measurement of diamond CSDs presents a number of unusual problems. Diamonds are very rare in most rocks; hence a large sample size is needed. Even then it is difficult to make sure that the sample is from a geologically homogeneous unit – mixed diamond populations in the sample are almost inevitable. In some deposits many diamond crystals are broken during emplacement, and it must be clearly established how broken crystals will be treated.

Commercially diamonds are separated by mechanical methods. Such a process can lead to two problems – yield variations and broken crystals. Although the yield of larger crystals is high, it may fall to zero for smaller, commercially unimportant diamonds (<1 mm). If the yield variations are known then compensation is possible, except where the yield is zero. As has been shown earlier, it is important to establish the lower size limit of detection. Upper size is also limited by the crushing process, which may break larger crystals. However, in some deposits natural forces have already broken many of the crystals during emplacement and these cannot always be distinguished from breakage during processing.

Exploration studies commonly use total solution methods, in addition to mechanical separation. A smaller sample is coarsely crushed and fused with alkali carbonates. A number of minerals remain and the diamonds are selected by hand from this concentrate. There is no lower limit to the size of diamonds that can be recovered. However, the smaller sample size of solution methods limits the largest crystals that can be recovered, as does the initial crushing of the sample.

The most extensive study of diamond sizes is that of Rombouts (1995). He considered that diamond sizes in most deposits follow a lognormal distribution. However, he pointed out that this commonly breaks down at extreme sizes. He notes that many diamond deposits have exponentially increasing crystal numbers versus diameter. This is what is observed in many other rocks (a 'straight' CSD). Unfortunately, there is little published data that can be used to verify this.

One study has been published on the sizes of diamonds in a lamproite from Australia (Figure 3.55; Deakin & Boxer, 1989). Diamonds were extracted in two ways: mechanical and solution. Few crystals were broken; hence a CSD can be constructed. The two curves can be fitted together, if allowance is made for the low yield of small stones in the mechanical separation. The resulting CSD is relatively straight from 0.15–3.00 mm, but with a slight turn-up for small crystals. This suggests that diamonds grew under conditions of increasing undercooling. Some nucleation and growth must have occurred just before eruption. There is no evidence for coarsening, but it may have been concealed by the last stage of growth.

Some eclogite xenoliths in kimberlites are very rich in diamonds (up to 0.5%). Several of these xenoliths have been studied by X-ray tomography (e.g. Taylor, 2000). However, no information has been published on the size distribution of diamonds in these rocks. It is true that the crystal numbers are low; typically less than 30, but even so there would be interest in such data.

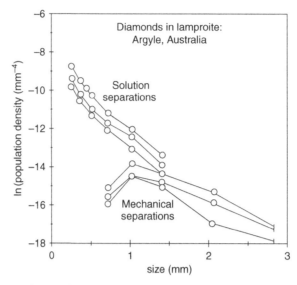

Figure 3.55   Diamonds in lamproite. Smaller diamonds were extracted by solution methods, where mechanical separations were used for larger diamonds. After Deakin and Boxer (1989).

3.4.5.2 Carbonates and chromites

Peterson (1990) studied the CSDs of several unusual minerals from carbonatite and other lavas of the 1963 eruption of Oldoinyo Lengai. He examined the alkali carbonate minerals gregoryite ($(Na_2,K_2,Ca)CO_3$) and nyererite ($Na_2Ca(CO_3)_2$) from carbonatite lavas and combeite ($Na_2Ca_2Si_3O_9$) from a nephelinite lava. In the original paper Peterson did not describe how he calculated the CSDs from the intersection data. However, in a later paper part of the earlier data was recalculated (Peterson, 1996). His revised data have straight CSD, right down to very small crystals. This is consistent with continuous nucleation and growth.

Waters and Boudreau (1996) examined chromite cumulates in the Stillwater complex. The chromites occur in centimetric layers. Even in thin section the crystals in the centre of the layer are clearly larger and more abundant. The CSDs confirm that the centre has been coarsened, compared to the edges. This behaviour is consistent with the DeHoff (1984) model of coarsening, as the growth rate will be higher where the crystal edges are closer together. This type of positive feedback, reinforcing early, weak structures has been explored by Boudreau (1987, and subsequent papers) and may account for the well-known 'inch-scale' layering in the Stillwater complex.

3.4.5.3 Chondrules and minerals in meteorites

Study of meteorites differs from most other aspects of geology as data from a small number of small samples are extrapolated to reveal the essential character of large bodies. So far textural studies have been concerned with three aspects of meteorite petrology: size distributions of chondrules in chondritic meteorites, CSDs of minerals in igneous-textured meteorites and CSDs of magnetic mineral grains possibly of biological origin.

Most silicate-dominated meteorites contain at least some chondrules and they may dominate the rock. Chondrules are spherical globules of silicate melts (~0.15–1 mm) that are present in almost all silicate meteorites. They were formed by flash heating of dust in the solar nebula and their size distribution has been used to infer the dynamics of the nebula (Rubin, 2000). Given the importance of chondrules, there have been a number of studies of their size distribution. In several early studies meteorites were disaggregated and the true size of the chondrules measured. However, not many meteorites are amenable to this treatment and information on the internal structure of the chondrule is not available. Many more studies have determined chondrule sizes from examination of sections. In many cases intersection data were not converted to 3-D values but examined directly (e.g. Rubin & Grossman, 1987). This is useful for showing differences in sizes between chondrules in different meteor classes, but conclusions based on fits of the size distribution to well-known models are weak. In other cases conversion equations derived empirically for sedimentary rocks were used (Friedman, 1958) that suggested that chondrules were actual smaller than their intersections, which is clearly wrong. Eisenhour (1996) proposed the use of a Saltikov-type method, modified to account for the thickness of the thin section. He found that chondrule size distributions that were formerly considered to be lognormal actually fitted more closely the Weibull distribution. This is important as lognormal distributions have sizes down to zero, whereas Weibull distributions can have a minimum size.

Shergottites are rare meteorites that are essentially basaltic rocks from Mars. Many shergottites have olivine, pigeonite and/or augite phenocrysts set in a fine-grained matrix dominated by plagioclase. CSDs of some of these minerals have been examined in four different studies (Lentz & McSween, 2000, Taylor *et al.*, 2002, Goodrich, 2003, Greshake *et al.*, 2004). In all these studies the CSDs were determined from intersection dimensions in thin sections and, where specified, converted using the inaccurate 'Wager' method. The lower size limit was not specified, but appears to be close to the thickness of a thin section, 0.03 mm. They found turn-downs for small crystals in some samples that may just be an artefact of this lower measurement limit. In addition, they used equal sized intervals, hence many larger size bins were empty. They then went on to connect up these isolated bins and concluded that some of the CSDs had two different slopes. While this may be correct, the data as presented in the papers do not prove this. Clearly this is an interesting subject for CSDs studies, but perhaps the CSDs of many shergottites should be studied together using more modern conversion and display methods.

There has been considerable interest in the possibility that some meteorites may contain fossil bacteria from Mars, hence the first signs of extraterrestrial life. One possible indicator of the presence of fossil bacteria is magnetite crystals which are used by bacteria on Earth to orientate themselves (McKay *et al.*, 1996). The bacterial magnetite crystals have distinctive shapes and size distributions, which could be used to identify Martian magnetofossils. Thomas-Keprta *et al.* (2000) separated magnetite crystals from carbonate globules in the Martian meteorite ALH84001 and examined them using a transmission electron microscope. They also extracted magnetite crystals from a bacterium for comparison. The magnetite crystals were sorted into different shapes and the size distributions determined for these projected images. Magnetite crystals in the bacterium were prismatic with a narrow range in size and shape. Thomas-Keprta *et al.* (2000) considered that the prismatic fraction of the meteoritic magnetite crystals also had a similar range in shape and size which suggested that they may also be of bacterial origin. There have been many criticisms of this conclusion, summarised by Treiman (2003). One concerns the size distribution of the meteoritic magnetite crystals: the size distribution appears to be much wider than that of the bacterial magnetite crystals. Treiman (2003) suggested that the magnetite formed by shock and thermal metamorphism of the carbonate. This could form the observed lognormal distribution.

3.4.5.4 *Hematite in pseudotachylytes*

Pseudotachylytes are melts that are produced along faults during movements: they both melt and solidify very rapidly. Petrik *et al.* (2003) have examined the CSDs of hematite that crystallised from pseudotachylyte melts. The authors converted intersection data in two different ways in different parts of the paper: where stereologically correct methods are used the CSDs are straight down to very small crystal sizes. However, elsewhere the authors used the 'Wager' conversion and interpreted the observed turn-down for small crystals as indicating coarsening or closed-system crystallisation. It is more likely that this feature is an artefact caused by both incomplete counting of small crystals and incorrect conversion of data. The hematite CSDs indicate crystallisation times of 10 seconds at the tips of the veins and up to 90 seconds within the wider parts of the veins. Growth rates of hematite are about five orders of magnitude faster than that of ilmenite in a lava lake.

3.4.5.5 Experimental studies

Although there has been much work in experimental petrology on the nature of phases under different pressure and temperature conditions, there have been few experimental studies that use the distribution of crystal sizes. Cabane *et al.* (2001) examined the kinetics of coarsening (Ostwald ripening) of quartz in simplified synthetic silicate melts. They did observe coarsening, but the kinetic factors did not match that predicted by LSW theory (Lifshitz & Slyozov, 1961). They suggested that this may be caused by an initial transient regime or that coarsening is not limited by diffusion, but by surface nucleation. However, it may also be related to the ideal nature of the experimental materials, or that the LSW model is not the best to apply to geological materials (DeHoff, 1991, Higgins, 1998). Their work suggests that coarsening may be very important just after nucleation, but that it is not important for quartz crystals larger than 1 mm during geologically reasonable periods of time. However, they admit that it may be a different story for other minerals and magmas.

3.4.6 *Undeformed metamorphic minerals*

3.4.6.1 Garnet

Garnet is by far the most popular mineral for CSD studies in metamorphic rocks, with the first study in 1963 by Galwey and Jones (1963). This may reflect the ease with which it can be separated from other minerals. Most studies have been on rock sections: the equant shape of most crystals minimises the stereo-logical cut-section effect, but the intersection-probability effect must still be considered (see Section 3.3.4.3). There have also been a number of studies in three dimensions: these enable the spatial relations between grains to be examined in detail. However, 3-D analysis is not always better than sectional analysis as resolution problems with some X-ray methods may limit the size range of crystals that can be examined.

Early studies established that garnets generally have a unimodal concave-down size distribution (Galwey & Jones, 1963, Galwey & Jones, 1966, Kretz, 1966a), although Kretz (1966a) did find one bimodal distribution. The precise shapes of the CSDs are not easy to establish from these studies, but appear to be between normal (Kretz, 1993) and lognormal (Cashman & Ferry,

1988) by population density (= frequency, if the size intervals are constant). X-ray tomographic studies have improved the quality of the CSDs by the elimination of stereological effects (Carlson *et al.*, 1995). In all studies there appear to be no small crystals; however, the lower size limit of detection is not generally well defined and hence this feature may be in some cases an artefact.

There has been much debate on the interpretation of garnet CSDs. The most popular theory is that garnet nucleated, grew and then became coarsened by Ostwald ripening following the LSW model (Cashman & Ferry, 1988, Miyazaki, 1996). Coarsening requires transport of material between crystals and this is easy if there is a fluid present (silicate or water dominated), or if the transport distance is short, as in monomineralic rocks. Carlson (1999) has argued that Ostwald ripening is severely limited by diffusion in polymineralic rocks under metamorphic conditions. He also points out that the shape of the garnet CSDs is not that expected for LSW coarsening. However, it has been pointed out that the LSW model may not be the most appropriate for geological materials (DeHoff, 1991, Higgins, 1998). In addition it is possible that deformation promotes circulation of fluids by opening up channels. An interesting twist on the problem is suggested by the data of Daniel and Spear (1999). They looked at Mn zoning patterns in garnet crystals and concluded that they were formed by the nucleation and growth of many crystals that subsequently rotated slightly and coalescenced.

3.4.6.2 Accessory minerals

Accessory minerals are very important for geochemical modelling and geochronology, but little quantitative work has been done on their textures, except for zircon (see Section 3.4.3.3). Zeh (2004) examined the CSDs of apatite, allanite and titanite in four gneiss samples that were metamorphosed to maximum temperatures of 550 °C to 680 °C. He used a variety of 3-D techniques to determine the CSDs of these minerals over a large size range. All CSDs are unimodal, concave down. On a classical CSD diagram the peak shifts to larger sizes as the slope of the right side fans around to shallower slopes. Zeh (2004) constructed a complicated growth history using the growth model of Eberl *et al.* (1998) followed by coarsening. However, the CSD patterns closely resemble the progression seen in other rocks that has been explained just by coarsening (e.g. Cashman & Ferry, 1988, Higgins, 1998), especially that following the 'Communicating Neighbours' equations of DeHoff (1991). Coarsening in metamorphic rocks may be ubiquitous; hence the nature of the original CSD is not observable.

3.4.6.3 *Crystals in migmatites*

Migmatites are rocks produced during partial melting. There are a number of aspects of interest here: the size-dependence of melting can be investigated, but is usually concealed by later processes. Berger and Roselle (2001) instead examined the CSD of cordierite produced by melt-producing reactions. They found that it was straight and proposed a nucleation and growth model similar to that for crystallisation of minerals from magma. Plagioclase in a leucosome had a concave down CSD suggesting coarsening. K-feldspar had a bimodal CSD that was not easy to explain. Berger and Roselle (2001) only examined three samples in all: clearly this material has a lot of promise for further study.

3.4.6.4 *Hydrothermally grown crystals*

Study of crystal sizes in hydrothermal crystallisation experiments by Eberl *et al.* (1998, 2002) led to their proposal that all CSDs are lognormal and that this necessitates a new growth theory, which they entitled 'the law of proportionate effect'. However, as has been seen above, there are many processes that can affect CSDs and it seems simplistic to force fit one model, especially when it has no experimental basis. In addition, some of the studies of Eberl and his coworkers have been hampered by incorrect conversion equations from intersection measurements to CSDs. Finally, the frequency diagram used in these studies is not very sensitive to variations in CSD shape. Bindeman (2003) has shown that some CSDs are indeed lognormal, but this is more likely produced by coarsening or repeated cycles of growth and solution (see Section 3.4.3.2).

3.4.7 *Deformed and broken crystals and blocks*

3.4.7.1 *Olivine in mantle lherzolites*

Mantle samples fall between the simple classification of igneous and metamorphic rocks – they are clearly igneous in origin, but have been deformed. There are few studies of the CSDs of minerals in mantle samples despite the volumetric dominance of the mantle in the Earth, the enormous number of fabric studies and the need for size data in geophysical modelling. Armienti and Tarquini (2002) examined the CSDs of olivine in suites of spinel and garnet lherzolites with a range of textures. They found that the CSDs followed a power law distribution over two orders of magnitude of size. Granoblastic textured samples had fractal dimensions around 2.4, which compares with values of 2.6 for fragmentation models (see Section 3.3.7.1). Porphyroclastic and cataclastic textured samples had fractal dimensions up to 3.8, but the interpretation of this is not yet clear.

3.4.7.2 Dynamic recrystallisation

The fabric of naturally deformed minerals has been studied extensively generally by using the mean grain size. Stipp and Tullis (2003) examined experimentally the relationship between mean grain size in quartzite and flow stress. They ensured that deformation reached equilibrium and that stress was measured accurately, but unfortunately did not apply stereological corrections to their grain sizes values. They found that mean grain size was well correlated with flow stress within the dislocation creep regimes. A slightly different correlation was observed at greater flow stresses.

Herwegh *et al.* (2005) have pointed out that the full grain size distribution is important in controlling creep behaviour, although it is rarely measured. In mylonites deformation of the rock reduces grain size at the same time as recrystallisation repairs the damage. The result is an equilibrium grain size distribution that depends on the temperature and the strain rate. Herwegh *et al.* (2005) examined the CSDs of calcite from mylonites in the Helvetic Alps. They found concave down frequency distributions, which they summarised using mean, mode, skewness and kurtosis. They showed that deformation follows grain-size-sensitive mechanisms, such as volume or grain boundary diffusion and grain boundary sliding.

It seems logical that the presence of water should increase the growth rate of minerals during deformation and recrystallisation. Experimental studies of olivine proved this to be true (Jung & Karato, 2001). The size of olivine subgrains was determined from their intersection lengths. These values were multiplied by 1.5 to estimate the 3-D size, but no correction was made for the intersection-probability effect. These modified intersection lengths had a lognormal distribution. Olivine deformed under dry conditions had a mean intersection size of 3 μm, whereas in the presence of water the mean intersection size was 15 μm.

3.4.7.3 Broken blocks and crystals

Rocks are fragmented during engineering and mining operations: hence, the size distribution of fragments has economic implications, such as in the processing of mining tailings. Turcotte (1992) has compiled a number of studies of the size distribution of broken rock (Figure 3.56). The data covered an unusually wide range of sizes – up to four orders of magnitude. He concluded that explosively crushed rock had a fractal dimension of about 2.5. This compares closely to a simple fragmentation model that produced a fractal dimension of 2.58.

A fractal model of fracturing has not been universally accepted: Suteanu *et al.* (2000) mention that some authors have considered that fracturing

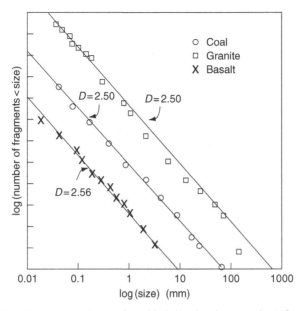

Figure 3.56 Fragment sizes of artificially broken rock (after Turcotte, 1992). Size is cube root of fragment volume. *D* is the fractal dimension.

produces multimodal distributions. Their experiments suggest that this may reflect the way data have been presented, however, they also offer the possibility that size distributions are multifractal – that is, the fractal dimension changes with the size. Data quality must be very high to be able to prove multifractal size distributions, but a change in fractal dimension is to be expected around the typical grain size of the rock.

Engineering geologists commonly have to deal with heterogeneous rocks that are comprised of hard blocks of variable size in a weaker matrix. Such 'block-in-matrix rocks' (bimrocks) would be described by geologists as conglomerates, breccias and fault gouges. However, an engineer is more interested in the mechanical properties than in the origin of the rock. Hence, it is important to be able to quantify the size distribution of the blocks in bimrocks. Medley (2002) has shown that this can be done from measurements of blocks intersected by drill holes or surface exposures using the same stereological methods that are used for examining crystal sizes.

Kimberlites are magmatic rocks that are commonly loaded with xenoliths, and their size distribution can be used to determine their mode of origin. Kotov and Berendsen (2002) examined smaller xenoliths in a kimberlite pipe from Kansas. They present few details of their results but conclude that the sizes of xenoliths follow a fractal distribution, here transformed to the Pareto distribution so that it could be tested statistically. This distribution is consistent with a model for the origin of the xenoliths by deep explosion without further breakage during transport. In half of the samples minor divergence from this distribution could be ascribed to Bagnold effect sorting.

Fault gouge is the powdery rock developed along faults during earthquakes. Measurement of the particle size distribution (PSD) of gouges is not easy as they are very fine grained and particles tend to aggregate. Wilson *et al.* (2005) used a laser particle-size analyser to determine PSD and found that the PSD varied over time as the aggregates broke down. However, at no time was the distribution fractal as has been noted for larger size fractions. They concluded that fault gouge did not form by wear and attrition of the fault surfaces, but by dynamic rock pulverisation during each earthquake. In addition, creation of the large surface area of the gouge particles could easily take a sizeable proportion of the energy of an earthquake.

Some crystals in volcanic rocks are broken (Allen & McPhie, 2003). So far study has been limited to the overall abundance of broken crystals, rather than their size distribution. In one ignimbrite 67%–95% of crystals were broken, but in intermediate and silicic lavas and tuffs a more typical value is 5%–10% (Allen & McPhie, 2003).

Notes

1. There has been debate in petrology about the use of the terms solution and melting. We insist that children are incorrect when they talk of sugar melting in water, but we commonly talk of granite melting to a silicate liquid. To a chemist a liquid must be a pure substance; hence in chemical terminology, both these examples are dissolution and silicate liquids are in fact high-temperature solutions. Hence, rocks do not melt, they dissolve. However, we all know what we mean by melt, even if we use different terms.
2. There is another terminology problem here. To a physicist a fluid is just a liquid, gas or super-critical fluid. However, many geologists use the term fluid if the substance is dominated by water, and melt if it is dominated by silicates (or carbonates, sulphides, etc.). In actual fact there is a continuum of fluids with compositions from pure water to pure silicate. And, as noted above, they are all actually solutions.
3. Also called the Schwartz–Saltykov method.

4

Grain shape

4.1 Introduction

Grain shape is something that is easy to express qualitatively, but difficult to quantify precisely (Costa & Cesar, 2001, Verrecchia, 2003). In earth sciences we do not generally need extremely precise measures of shape, because the objects we deal with are commonly not perfectly regular. However, quantification of aspects of shape can help in understanding the petrogenesis of rocks. The subject naturally falls into two domains. For crystalline rocks (igneous and metamorphic rocks, chemical sediments and hydrothermal ore deposits) we are concerned with the shape or habit[1] of crystals, whereas in clastic rocks we want to quantify the shape of clasts and grains. The methodology of these two fields is partly shared but could benefit from more exchange.

The overall shape of crystals reflects growth, solution and deformation. Crystals that grew unimpeded from a fluid may have many different forms, but all are ultimately controlled by the crystal structure. We usually think that faces bounding such grains should be flat, but in some minerals growth faces are curved. In thin section crystal outlines are qualitatively classified as euhedral, subhedral or anhedral.

The habit of euhedral crystals has been studied for a long time (see the introduction in Sunagawa, 1987a): in 1669 Nicolaus Steno proposed that the interfacial angles of a crystal were constant and that the external form of a crystal growing in a fluid depends on the relative growth rates of the different faces. In many mineralogical and petrological studies the habit of crystals has its own interest (e.g. Sunagawa, 1987b). For instance, it is commonly observed that crystal habit varies with growth environment and size (see the review in Kostov & Kostov, 1999): plagioclase crystals in the groundmass of lava are tabular (commonly misnamed 'lath-shaped') whereas those in more slowly cooled rocks are more equant. Rapid growth forms, such as spherulites, dendrites and swallowtails, are not treated here as they are difficult to quantify.

Crystal shapes can help in the quantification of other textural parameters: an estimate of shape is needed to carry out some types of stereological corrections (e.g. Higgins, 2000). For this purpose overall crystal shapes are simplified and defined in terms of the aspect ratio and degree of rounding. This simplified scheme can also be used to examine crystal shape variations between different petrologic environments.

Grain boundaries and dihedral angles are other quantifiable aspects of crystal shape: grain boundaries may be flat, curved or sutured, and all these can be quantified. Dihedral angles are the angles at the corners of crystals (Smith, 1948). It has been shown that the values of such angles are a measure of the textural maturity of a rock (Elliott *et al.*, 1997).

4.2 Brief review of theory

4.2.1 Shapes of isolated undeformed crystals

4.2.1.1 Equilibrium forms

Crystal growth and final habit are controlled by many factors (Baronnet, 1984, Sunagawa, 1987a): a good point of departure is the equilibrium form of the crystal. This form is purely theoretical and is that in which the surface energy (or surface tension) of the crystal is minimised (Wulff, 1901, Herring, 1951a). Unlike a liquid, the surface energy of a crystal varies with orientation; hence the equilibrium form of a crystal is never a sphere. The energy associated with a crystal face depends on the density of lattice points – the reticular density. Lattice points are not necessarily atoms, but points in space that have the identical environment in the same orientation. A crystal face is more developed if it has a high reticular density; that is it intersects a large number of lattice points (Bravais law). A more recent theory uses the concept of periodic bond chains (PBC; Baronnet, 1984). A PBC is an uninterrupted chain of strong bonds between atoms. On this basis faces can be classified: F (flat) faces have two or more PBCs, S (stepped) faces have one PBC, and K (kinked) faces have no PBCs at all (Figure 4.1). It may seem counter-intuitive, but the largest faces of a crystal have the lowest growth rates. The F faces grow by layer mechanisms and hence grow at the slowest rates. Therefore, the facets of a crystal are generally the F faces.

The equilibrium form of a crystal is controlled by the crystalline structure and hence does not depend on the environment in which the crystal grew. However, it does depend on extensive variables, such as temperature. At

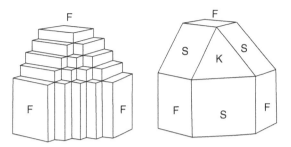

Figure 4.1 Periodic bond chains and crystal structure. F = flat faces; S = stepped faces; K = kinked faces.

higher temperatures the energy difference between different faces becomes less and it becomes energetically feasible to have curved faces. At the extreme this tends to a sphere with minor facets. The equilibrium forms of common minerals broadly resemble the observed forms in the expression of faces, but the importance (relative size) of the faces does not always agree.

4.2.1.2 Growth forms

There are many different growth forms of crystals, even under identical thermodynamic conditions (Sunagawa, 1987a). The broad classification of crystals whose growth has not been impeded by the presence of other materials is into spherulitic, dendritic, hopper and polygonal. The different faces of polygonal crystals may be developed to different extents, giving different polygonal habits. Where crystals are growing in the presence of a second phase they may be incorporated into the second phase as chadocrysts, or they may stop growth, leading to anhedral (irregular) forms. Some of the latter may have curved faces formed by the simultaneous growth of two crystals.

The growth habit of crystals is fundamentally controlled by growth rates of the various faces or directions. Such growth rates are controlled by intensive parameters such as temperature, pressure and chemical potential gradient (Figure 4.2; Kouchi *et al.*, 1986, Kostov & Kostov, 1999). Where crystals are surrounded by other crystals, for instance in the sub-solidus, other parameters are important, most significantly the interfacial energy (see dihedral angles below; Hunter, 1987).

4.2.1.3 Solution forms

The solution forms of crystals have not been studied in as much detail as growth forms. A common observation of solution forms is within crystals, as revealed by zoning. For instance, the optical properties of plagioclase make

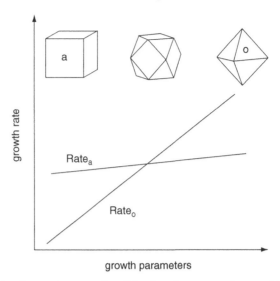

growth parameters

Figure 4.2   A schematic example of the development of crystal forms related to different growth rates under different growth parameters. The a and o faces have growth rates that vary in a different way to growth driving force parameters, such as undercooling (supersaturation), chemical environment or composition.

it easy to observe minor variations in composition. Resorption surfaces that cut across fine composition zones are commonly encountered, especially in volcanic rocks (e.g. Stamatelopoulou-Seymour *et al.*, 1990). In all cases the solution form is rounded. That is, the radius of the corner is increased progressively by the consumption of the faces. The solution surface is frequently rough, reflecting the effects of dislocations and impurities in the crystal. Other minerals, such as quartz, pyroxene and olivine have resorbtion surface zones similar to those of plagioclase, but they are not so easily seen, except with techniques like Nomarski microscopy (see Section 2.5.3.3). Most natural diamond crystals have suffered solution during transport from the mantle, which is expressed by curved crystal surfaces. The process starts with loss of the sharpness of the edges and proceeds to the faces, which become curved (Sunagawa, 1987b). Extensive solution leads to rounded crystals.

4.2.2 Shapes of deformed crystals

The shape of deformed crystals, like their size, is a balance between the processes that deform crystals (see Section 3.2.5), which give the crystal an excess of energy, and the processes that will dissipate that energy by the elimination of lattice defects (relaxation) and the reduction of surface area

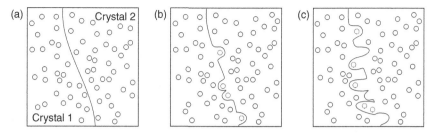

Figure 4.3 Schematic development of sutured grain boundaries between two crystals. (a) Initial state after deformation. Circles are lattice defects. (b) Bulging of the crystal boundary removes a defect from the adjoining crystal. The new boundary is then close enough to capture other defects. (c) The process will continue while it is energetically favourable to make more grain surface and lose lattice defects.

per unit volume. In deformed rocks the energy associated with lattice defects is typically two to four orders of magnitude greater than that associated with excess surface area (e.g. Duyster & Stockhert, 2001). During uniform deformation equilibrium may be reached between these processes, giving a constant crystal shape that depends on the deformation style and rate, and the temperature and fluid content. Of course, such information may be masked by subsequent deformation under different conditions. Deformation mechanisms have been discussed briefly in Section 3.2.5. Recovery of surface energy by coarsening (ripening) has also been discussed with reference to crystal size (see Section 3.2.4). However, the role of lattice defects has yet to be discussed.

Lattice defects have excess energy compared to other parts of the crystal lattice, hence will be removed during relaxation. Recovery of such lattice defects close to the edge of a crystal can occur by grain boundary migration (Kruhl & Nega, 1996). A suitably located defect can be eliminated by removal of that part of the grain and reconstitution of the atoms into the adjacent grain (Figure 4.3). The overall effect is a migration of the grain boundary and the bulging of one grain into another, leading to a sutured grain boundary. The energy expended in creating a larger grain surface area is less than the energy recovered from the elimination of the defect.

4.2.3 Dihedral angles of crystals in rocks

So far the shape of isolated crystals has been discussed, but in most rocks crystals touch. If such a material is completely solid then four crystals meet at most crystal corners and three meet along most crystal edges. In the plane normal to the edge the angle of the traces of the crystal boundaries is the dihedral angle (DA).

Crystals will grow unimpeded in magma or liquid until they impinge another crystal (Figure 4.4). The impingement DAs for such textures are very variable but the population of DAs has a mean of 60° (Figure 4.5). This texture is regarded as the departure point for further textural development in solidifying systems (Holness *et al.*, 2005). Such textures should be developed in simple magmatic systems without magma flow or imposed stress. They may also develop in some hydrothermal ore deposits and chemical sediments.

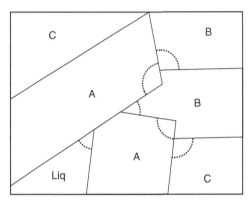

Figure 4.4   Impingement angles in growth textures. Here, three different phases (A, B and C) have grown from a liquid without equilibration and a portion of the liquid remains (Liq). Some of the many dihedral angles associated with each phase are shown by arcs. In a simple growth texture like this these dihedral angles are referred to as impingement angles.

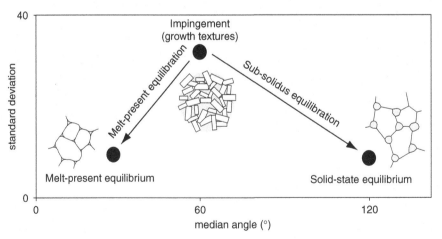

Figure 4.5   Changes of the median and standard deviation of dihedral angles from initial impingement (growth) textures in response to melt-present and sub-solidus equilibration (from Holness *et al.*, 2005).

At equilibrium the DA reflects the difference in surface energy (interfacial tension) between the phases. If three phases are isotropic and meet at an edge, then (Smith, 1948, Smith, 1964)

$$\frac{\gamma_1}{\sin \alpha_1} = \frac{\gamma_2}{\sin \alpha_2} = \frac{\gamma_3}{\sin \alpha_3}$$

where γ_1, etc. are the surface energies between the phases along a grain boundary, and α_1, etc. are the DA between the phases on the opposite side of the edge (Figure 4.6a). Liquids are isotropic, but all minerals are physically anisotropic, hence the surface energy varies with direction and even at equilibrium there will be a range of DAs (Herring, 1951b, Laporte & Provost, 2000). Experimental work suggests that the degree of anisotropy broadly correlates with the overall symmetry of the minerals: it is lowest in isometric minerals like spinel and highest in monoclinic minerals like diopside (Laporte & Watson, 1995, Schafer & Foley, 2002). Some studies have shown that true F-faces (Figure 4.1) may form where one of the phases is a fluid, leading to further complexity in the anisotropy of surface energy (Cmiral *et al.*, 1998, Kruhl & Peternell, 2002) although this may not always be the case (Hiraga *et al.*, 2001). In a polymineralic rock many different mineral combinations are possible, each with their own range of equilibrium DAs (Figures 4.6b, c, d).

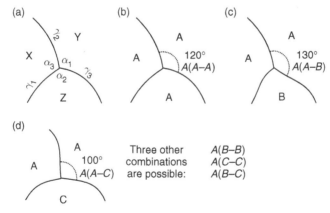

Figure 4.6 Equilibrium dihedral angles in a three-phase system. (a) If three different isotropic phases X, Y, Z, meet at an edge then the DAs α_1, α_2, α_3 are simply related to the interfacial energies of the grain boundaries γ_1, γ_2 and γ_3. (b) If the rock is monomineralic then the mean equilibrium DA will be 120°. The DAs will be variable because the surface energy of minerals is not isotropic. (c, d) Here, the rock has three phases A, B and C, one of which may be a liquid. In this case the equilibrium DA for each mineral varies with the nature of its neighbours, as well as the anisotropy. In a three-phase rock, six different combinations are possible, of which three are shown.

Clearly, surface energies are important in the development of texture and their values should be determined. Such values may be calculated for crystals in a vacuum, an uninteresting value for most geological studies. DAs can provide useful estimates of the ratios of surface energies (Kretz, 1966b, Vernon, 1968, Hiraga *et al.*, 2002), but it is much more complicated to measure the actual surface energies. Surveys in materials science suggest that values are of the order of $1\,\mathrm{Jm}^{-2}$ for many contacts between crystalline materials (Sutton & Balluffi, 1996). The surface energy of olivine–olivine contacts was determined from a mantle sample to have a maximum value of $1.4\,\mathrm{Jm}^{-2}$ for the greatest degree of crystal misorientation (Duyster & Stockhert, 2001). Surface energies between solids and liquids are much lower than these values, because the liquid is structurally isotropic. Mungall and Su (2005) found that the surface energy between sulphide and silicate liquids was 0.5–$0.6\,\mathrm{Jm}^{-2}$. They note that the surface energy of silicate mineral–silicate liquid contacts should be much lower.

Dihedral angles can be used as a measure of the degree of textural equilibrium in a rock (Elliott & Cheadle, 1997, Holness *et al.*, 2005). A rock is in textural equilibrium if 'the surface topology of the grains is in mechanical and thermodynamic equilibrium' (Elliott *et al.*, 1997). If a rock has coarsened in a stress-free environment then it will have approached textural equilibrium (see Section 3.2.4). The attainment of textural equilibrium can be assessed from the distribution of grain sizes, from the curvature of the grains, or from the distribution of DAs. In general DAs are the most sensitive to equilibration and grain shape and size the last to respond (Laporte & Provost, 2000). In a rock composed of only one phase the mean DA is $120°$ (Figure 4.6b). If several phases are in equilibrium then the DA for a particular mineral will depend on its neighbours (Figures 4.6c, d) in addition to the anisotropy of surface energy (Kretz, 1966b, Vernon, 1968, Elliott *et al.*, 1997, Holness *et al.*, 2005).

DAs are of particular interest where one of the phases is a liquid, which is isotropic. In this case the DA is a measure of the wetability of the surface by the liquid: if we consider again our ideal rock, composed of one isotropic mineral, then if the DA between the solid phase and a liquid is less than $60°$ then the liquid will form an interconnected network along three-phase junctions. This means that the connectivity threshold of the melt is zero. A survey by Smith (1964) suggests that this is the case for some minerals, aggregates of which are permeable even at very low degrees of melting. Conversely, if the DA is greater than $60°$ then the liquid will aggregate at the intersections of grains, and the connectivity threshold is much higher.

Calculation of rates of textural equilibration or coarsening requires absolute values of surface energies, which cannot be determined from DAs. However,

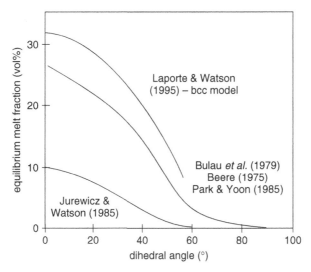

Figure 4.7 Minimum-energy melt fraction in partially molten rocks at equilibrium against dihedral angle (modified from Beere, 1975, Bulau *et al.*, 1979, Jurewicz & Watson, 1985, Park & Yoon, 1985, Laporte & Watson, 1995, Ikeda *et al.*, 2002).

approximate values estimated from an assessment of relevant experimental studies (Sutton & Balluffi, 1996, Holness & Siklos, 2000), suggest that textural equilibrium over a scale of millimetres could be achieved in partially molten silicate rock over tens of years.

A number of experimental studies have suggested that at equilibrium there is a minimum-energy melt fraction (MEMF) in a partially molten rock that depends on the DA (Figure 4.7; Jurewicz & Watson, 1985, Ikeda *et al.*, 2002). If the actual amount of melt is less than the equilibrium melt fraction then melt will be drawn into the crystal aggregate. However, if the material contains more melt than the equilibrium fraction, then the melt will be expelled and segregate outside the aggregate. These effects could cause the commonly observed clustering of crystals.

4.3 Methodology

4.3.1 Three-dimensional analytical methods

Grain shape is a property of the surface of the grain. It can be examined precisely using three-dimensional methods, if the grain surface can be separated perfectly from adjoining grains by physical or analytical means. This is the best way to look at the development of faces in minerals, and complex shapes such as dendrites. It is also necessary if there is a significant range in grain shape within a sample, for example for different grain sizes. The following 3-D methods have been used:

- The shape of individual grains can be determined by serial sectioning (see Section 2.2.1).
- In transparent materials, such as volcanic glasses, the shape of crystals can be measured directly with a regular or confocal microscope (see Section 2.2.2).
- X-ray tomography can be used to determine the shape of grains if adjoining grains can be separated (see Section 2.2.3).
- The shape of grains is preserved if they can be extracted intact from rocks by mechanical disintegration or partial solution (see Sections 2.5 and 2.6). Natural weathering can sometimes lead to the extraction of measurable grains (Higgins, 1999).
- Rocks commonly fracture along zones of weakness, which may include crystal surfaces or schistosity in clasts. Hence, aspects of the shape of individual grains can be commonly measured from such special sections exposed on irregular fractured blocks. This is particularly useful for tabular grains.

4.3.2 *Section and surface analytical methods*

It is not always possible to separate grains either chemically, mechanically or analytically for 3-D analysis. In these cases two-dimensional intersection or projection methods may give better results. Larger objects may be sectioned individually along special directions: this is important for looking at internal surfaces. However, generally the outlines of many grains are examined statistically and the mean shape determined. These methods may not be applicable if a range of shapes is present.

The shape of grains can be determined from images acquired from rock surfaces, slabs and sections, using a variety of techniques described in Section 2.5. Such images are processed manually or automatically to produce classified images (see Section 2.6). Finally, the classified images are processed to extract grain outline parameters. If the section is examined using techniques that can reveal the orientation of the crystal lattice, such as polarised light microscopy or EBSD, then the relationship between the shape and the lattice orientation can be determined.

Commonly data from many grains with different orientations are combined and the results interpreted in terms of grain shape. However, some shape parameters are difficult to measure by these methods and it may be better to examine grains with special orientations that are parallel or normal to the plane of the section. For crystals such orientations may be identified using optical methods such as birefringence and twinning. A universal stage can also be used to rotate crystals into the useful orientations (see Section 5.5.1).

Changes in the shape of a crystal during its lifetime may be observed from zoning patterns, or surfaces rich in inclusions inside the crystal (Sempels, 1978, Vavra, 1993). Such patterns are best revealed by regular optical methods, backscattered electron images (see Section 2.5.2.1), cathodoluminescence (see Section 2.5.2.3), Nomarski interference (DIC; see Section 2.5.3.3) or laser inferometry methods (Pearce *et al.*, 1987). Of course, important solution events may remove earlier history, so the story may be fragmentary. For quantitative studies of changes in morphology it may be necessary to use sections with special orientations (Pearce *et al.*, 1987, Sturm, 2004).

4.3.3 Extraction of overall grain shape parameters

4.3.3.1 Overall axial ratios

Grains can have very complex shapes but for some purposes all that is needed, or all that can be currently determined, are the overall shape in terms of its aspect ratio. There are a number of different definitions (see Figure 3.26).

- The dimensions of the smallest parallelepiped that encloses the grain. The aspect ratio then has three parameters: the short, intermediate, and long dimensions (S : I : L).
- The dimensions of a triaxial ellipsoid with the same moments of inertia as the grain. The aspect ratio is then the ratio of the minor (short) : intermediate : major (long) axes (S : I : L).

The aspect ratio of either model can be precisely determined from three-dimensional data for each grain. The limitations of three-dimensional methods have been described above; hence it may be necessary to use two-dimensional data from intersections or projections. For practical reasons it is easier to divide the total aspect ratio into the ratios S/I and I/L, both of which vary from 0 to 1.

In sedimentology the SIL values are commonly reduced to a single parameter, the sphericity:

$$sphericity = \sqrt[3]{(IS/L^2)}$$

Sphericity has a value of one for a sphere and decreases towards zero for more elongate clasts.

4.3.3.2 Intersection data

If grain shapes are convex, constant in shape and simple then parts of the overall aspect ratio may be estimated statistically from grain intersections using the distribution of the ratio of intersection widths and lengths (w/l), supplemented with other data where necessary. Obviously such methods cannot be used for concave or branching crystal forms, which must be measured by 3-D methods such as X-ray tomography or serial sectioning (see Section 2.2).

One simple method of estimating S : I : L is derived from a model of grains as parallelepipeds (Higgins, 1994). For tablet-shaped grains with no preferred orientation (massive) the mode of the w/l is equal to S/I (Figure 4.8a). For prisms the results are not so easy to interpret: w/l ratios have a wide distribution with only a minor peak at S/L (Figure 4.8b). A model based on ellipsoids

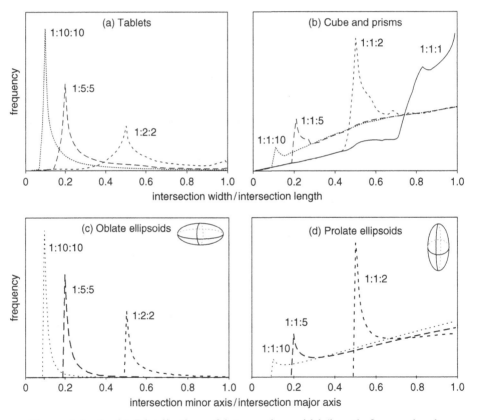

Figure 4.8   (a, b) Distribution of intersection width/length for randomly oriented planes intersecting a parallelepiped (after Higgins, 1994). Intersection dimensions are those of the smallest rectangle that can be fitted around the intersection. Representative crystal intersections are shown in Figure 3.27. (c) Intersections of a randomly oriented plane with an oblate ellipsoid (disc shaped; this study). (d) Intersections of a randomly oriented plane with a prolate ellipsoid (rugby/American football shaped; this study).

Table 4.1   *Modal values of different intersection parameters for populations of parallelepipeds with different fabrics (Higgins, 1994).* S = *short dimension,* I = *intermediate dimension,* L = *long dimension. For some orientations of strong fabrics values are invariant: these are shown in bold italic type.*

Fabric type and section orientation	Mode of intersection length (l)	Mode of intersection width (w)	Mode of intersection width/length (w/l)
Massive	I	S	S/I
Foliated – normal	I	S	S/I
Foliated – parallel	*L*	*I*	*I/L*
Lineated – normal	*I*	*S*	*S/I*
Lineated – parallel	L	I	I/L
F and L – normal	*I*	*S*	*S/I*
F and L – parallel	*L*	*I*	*I/L*

rather than parallelepipeds has a broadly similar distribution of intersection minor axis/intersection major axis ratios, equivalent to w/l. For oblate ellipsoids the ratio S/I (ellipsoid minor axis/ellipsoid intermediate axis) can be determined from the mode of the intersection minor axis/intersection major axis ratios (Figure 4.8c). As for prisms, the shape of prolate ellipsoids cannot be determined from the distributions of the intersection minor axis/intersection major axis ratios (Figure 4.8d).

Rocks with strong fabrics may have different w/l modal values depending on the orientation of the section with respect to the foliation or lineation (Table 4.1).

If a sample has a strong foliation or lineation and a section parallel to the fabric is available than the ratio I/L can be determined also. However, more commonly the ratio I/L must be determined from the distribution of the w/l ratios. Higgins (1994) showed that for parallelepipeds this ratio can be estimated from the w/l distributions as follows:

$$I/L = 0.5 + (\text{mean } w/l - \text{mode } w/l)/\text{standard deviation } w/l$$

However, this equation does not give accurate results for near equant shapes.

Garrido *et al.* (2001) did a very similar modelling of parallelepipeds and found a graphical relationship between the mean w/l, mode w/l and I/L. The following equation was derived from their graph.

$$I/L = (\text{mean } w/l + 1.09(\text{mode } w/l)^2 - 1.49(\text{mode } w/l) + 0.056)/0.126$$

Both methods need an accurate estimate of the mode w/l which is not easily available for most measured CSDs. In many cases it is best to measure I/L from crystals with special orientations (see Sections 4.3.1 and 4.3.2).

As was mentioned before the shape of prismatic crystals is not easily determined from w/l distributions. The best way to determine the shape of prismatic crystals is from the examination of crystals with special orientations: that is, with the long axis parallel to the section.

Some stereological methods for calculating crystal size distributions (e.g. CSDCorrections; Higgins, 2000) require also an estimate of the degree of rounding of crystals. For the models described above this parameter just modifies the shape model used for the conversion by combining intersections from parallelepipeds and ellipsoids.

4.3.3.3 Projection data

In some situations grain outlines are seen in projection and not in section. For instance, grains may be completely contained within a thin section. The projected outline width/length (w/l) can be used to determine aspects of the overall aspect ratio, as for intersection data, but the relationships are different. The parallelepiped model assumes that blocks are randomly oriented. Tablets have broad distributions of projected w/l that cannot be used to determine the aspect ratio of the grains (Figure 4.9a). However, prisms have well defined w/l peaks for which the mode = I/L (Figure 4.9b). For shapes between tablets and prisms the S/I ratio is not easily accessible and is best determined from grains with special orientations.

A model based on randomly oriented ellipsoid grains gives similar, but not identical results to the parallelepiped model: for oblate ellipsoids the modal projected minor axis/projected major axis has a peak at I/S, but it is narrow, with few intersections (Figure 4.9c). Hence, in practice it is not possible to determine the value of I/S for such shapes. However, for prolate ellipsoids the peak at I/L is well defined and the mode is easily measurable in real situations (Figure 4.9d). The I/S ratio of triaxial ellipsoids is not easily accessible from projection data.

If grains are deposited as a layer just one grain deep then their outlines can be easily imaged and measured. Here, the grains are not randomly oriented, but lie with the I–L plane horizontal. In this situation, the value of the w/l will be exactly equal to I/L, however S/I is not accessible from the image. If all grains are assumed to have the same composition then the total mass of the sample can be used to estimate the value of S/I (Mora & Kwan, 2000).

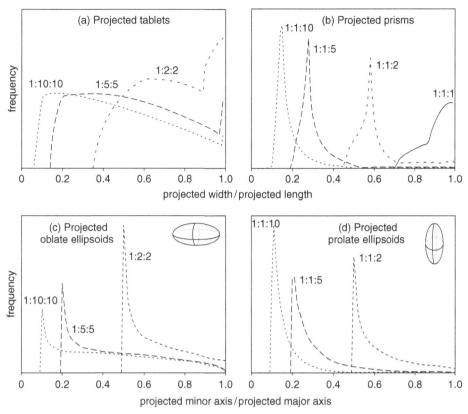

Figure 4.9 (a, b) Parallelepiped model of randomly oriented grains in projection (this study). The projected length and width are measured as the dimensions of the smallest rectangle that can be placed around the outline. (c, d) Ellipsoid model of randomly oriented grains in projection (this study).

4.3.4 Extraction of grain boundary parameters

The grain boundaries of crystals with curved, complex or sutured surfaces can be parametrised in a number of ways. Although grain boundaries are three-dimensional structures, so far they have only been investigated in detail in 2-D sections. The stereology of conversion of such 2-D measurements to 3-D parameters has not generally been investigated. The simplest descriptor of the shape of the grain is the 'Haywood circular parameter' which is the ratio of the actual perimeter to that of a circle of equal area. This parameter is more sensitive to the overall axial ratio of the grains (see Section 4.3.3) rather than the irregularities in the outline, which we seek here. The shape of loose rock particles is very important in sedimentology and hence the methodology is well established, but outside the scope of this volume (e.g. Barrett, 1980, Lewis & McConchie, 1994a, Lewis & McConchie, 1994b).

4.3.4.1 Fractal analysis of grain shapes

The shape of a grain boundary can be characterised by fractal methods (Mandelbrot, 1982, Orford & Whalley, 1983, Kruhl & Nega, 1996, Carey *et al.*, 2000). So far such methods have only been applied to sections or projections of clasts with the assumption that such measurements also characterise the parent 3-D object. A smooth surface (in a section) will have a fractal dimension of one, which will increase to a maximum of two for a plane-filling line. Outlines of particles of geological interest tend to have more restrained fractal dimensions from 1 to 1.36. However, it should be noted that this fractal dimension is that of a section through a surface, which will have a higher fractal dimension of between 2 and 3.

The fractal dimension of an outline is generally determined using the divider or box counting methods (Figure 4.10) and the results presented on Mandelbrot–Richardson (log–log) diagram. A single straight line on an M–R diagram indicates monofractal behaviour. That is, the shape is self-similar at all of the scales measured. Many objects of geological interest give multiple straight-line segments on an M–R diagram, which is termed multifractal behaviour. Some authors have distinguished two scales and others have divided up the fractal into several size segments (e.g. 3–30 pixels; 31–60 pixels; etc.), that were then analysed using multivariate statistics (Dellino & Liotino, 2002). Curved distributions on M–R diagrams indicate nonfractal behaviour and should not be forced into line segments. Care must be taken when outlines

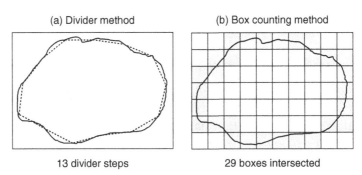

Figure 4.10   Methods for determining the fractal dimension of the outline of a grain. (a) Divider method. A line of length l is walked around the outline and the number of steps counted. This is repeated for different lengths and a graph of $\log(n)$ against $\log(l)$ is plotted (Mandelbrot–Richardson diagram). The slope is the fractal dimension of the line. (b) Box counting method. A box size of length l is defined and the fraction of boxes, n, that contain a part of the outline are counted. This is repeated for different box sizes and a graph plotted of $\log(n)$ against $\log(l)$. The slope is the fractal dimension.

are measured: the resolution, that is number of pixels, provides a lower size limit for the method. Also in many outlines the undulations are sufficiently infrequent that the starting position for the divider method affects the results. Several different positions and directions can be averaged to improve the resolution (Dellino & Liotino, 2002).

4.3.4.2 Gaussian diffusion–resample method

Saiki *et al.* (1997, 2003) have proposed a new rounding parameter that is not dissimilar to fractal methods. A selection of grains of similar size is first reduced to a single binary image (black and white) and the total area noted. The image is then passed through a Gaussian diffusion filter with an aperture of 0.5 pixels. The outlines are now blurred. The image is then rebinarised with a threshold such that the area of the grains is unchanged. The original and final images are subtracted so that the material eroded by the process can be measured. The process is repeated with a wider Gaussian filter. The eroded areas are normalised to the total area and plotted against the aperture: the slope of the line is a measure of the irregularity of the outline.

4.3.4.3 Fourier analysis of grain shapes

The overall shape of a grain intersection or outline can be evaluated precisely at all scales by the use of Fourier analysis (Ehrlich & Weinberg, 1970). However, data analysis methods may restrict the scale of application: most individual applications of a method cannot resolve the outline at a scale better than 0.1 to 0.01 times the size of the outline. More detailed measurements may be 'stitched together' from several analyses to overcome this problem. It should be also remembered that Fourier analysis is generally applied to only a single section through a three-dimensional object. Three-dimensional Fourier analysis is possible, but has not yet been done on geological materials.

The radius of the grain outline at angle Θ, $R(\Theta)$ is traced out. These data are transformed by classical Fourier analysis into a series of harmonics (Figure 4.11). The basic shape is a circle of equivalent area; the first harmonic is an ellipse and gives the elongation. The second harmonic gives a measure of triangularity and the third gives squareness. The harmonics continue up to the number of data points. The shape can be reconstructed as:

$$R(\Theta) = a_0 + \sum_{n=1}^{N} (a_n \cos n\Theta + b_n \sin n\Theta)$$

where N = the total number of harmonics, n = harmonic number, and a and b are coefficients that give the magnitude and phase of each harmonic. The

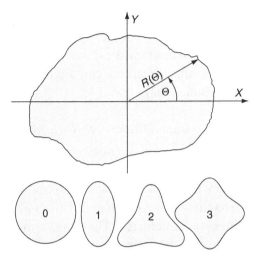

Figure 4.11　In classical Fourier analysis the radius of the outline, R, is decomposed into a basic circle ($n = 0$) and a series of harmonics, $1, 2, 3, \ldots, n$.

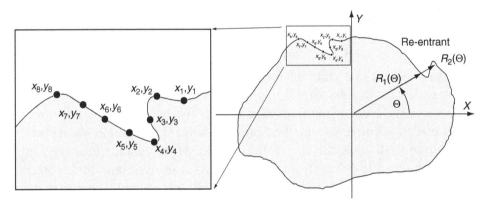

Figure 4.12　Re-entrant problem in classical Fourier analysis.

method is useful because it gives a measure of the shape of the profile at all scales, covering the field of sphericity, roundness and grain boundary shape (surface texture). There is, however, a problem for grains with re-entrants: there can be two radiuses for certain angles (Figure 4.12). This is not usually a problem for mature sediments, but occurs in volcanic fragments and grains in plutonic or volcanic rocks.

An approach that avoids this problem is the Fourier Descriptors method (Thomas *et al.*, 1995, Bowman *et al.*, 2001). In essence Fourier analysis is applied to the x, y coordinates of the outline of the grain and not its radius (Figure 4.12). The method uses complex numbers, with the x coordinate as the real part of a number and the y coordinate as the imaginary part. The outline is sampled at regular intervals to give a series of points x_m, y_m. The outline can be reconstructed from:

$$x_m + iy_m = \sum_{n=-N/2+1}^{+N/2} (a_n + ib_n) \left[\cos\left(\frac{2\Pi nm}{M}\right) + i\sin\left(\frac{2\Pi nm}{M}\right) \right]$$

where N is the total number of descriptors, n is the descriptor number, M is the total number of points, m is the index number of each point and a and b are coefficients for each descriptor (i denotes the imaginary part of a complex number). The descriptors $n = 0$, -1, -2, -3 give the radius (equivalent to size), elongation, triangularity and squareness as in classical Fourier analysis. The descriptors $+1$, $+2$, $+3$ give asymmetry, second-order triangularity, second-order squareness (Bowman *et al.*, 2001). The descriptors give almost too much information on shape. We are generally concerned more with the distribution of roughness at different scales. Drolon *et al.* (2003) have proposed that the Fourier descriptors are recast using wavelet techniques into a series of multiscale roughness descriptors. These give a measure of the total 'energy' at each scale. They appear to offer a simpler way of looking at roughness that is better at classifying sedimentary grains.

4.3.4.4 Shape parameters used in sedimentology

Some studies of crystalline rocks have used sedimentological shape parameters. One such measure is the circularity. This is defined by the perimeter of the grain outline:

$$circularity = 4\Pi(A/P^2)$$

where A is the area and P is the perimeter of the grain outline. It varies from one for a perfect circle to zero for an infinitely elongated grain outline. The elongation of a grain has several definitions: in 3-D it is the ratio of the longest and shortest axes. In 2-D it is the aspect ratio of its outline: that is, intersection length/width.

The grain shape can also be evaluated at a smaller scale by quantifying the roundness. This is usually measured from grain intersections using the Waddell formula:

$$roundness = \Sigma(r/R)/N$$

where r is the radius at the grain corners, R is the largest inscribed circle (circle enclosed in the grain outline) and N is the number of corners (Figure 4.13). This formula is rather difficult to apply to real grains; hence roundness is estimated from silhouette diagrams.

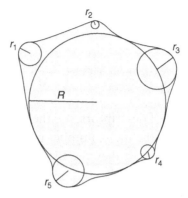

Figure 4.13   The 'roundness' parameter of grains, as determined from a section through the centre of a grain.

4.3.5 Determination of dihedral angles

Dihedral angles are a 3-D property and can be measured directly in transparent materials using a universal stage (see Section 5.5.1). The sample is rotated so that all three grain boundaries are vertical and the dihedral angles measured directly (Vernon, 1970) or using a stereographic plot (Vernon, 1970). It is also possible to use the tilting capabilities of a transmission electron microscope to examine true dihedral angles in thin films (Cmiral *et al.*, 1998). However, in most studies the apparent dihedral angles are measured in 2-D sections, such as thin sections and plotted on cumulative frequency diagrams. If there is only one true dihedral angle in a population then it is close to the median of apparent dihedral angles (Riegger & van Vlack, 1960). However, if there are a range of true dihedral angles, then this range cannot be determined uniquely from the population of apparent (2-D) dihedral angles (Jurewicz & Jurewicz, 1986, Elliott *et al.*, 1997). The overall distribution of apparent dihedral angles has been used to determine if a monomineralic rock is equilibrated or not (Elliott *et al.*, 1997). Impingement textures can clearly be distinguished from equilibrium textures (Figure 4.5), but the method is not sensitive to small variations in the degree of equilibration. Hence, it is better to use 3-D measurements where possible.

4.4 Typical applications

4.4.1 Undeformed crystalline rocks

The development of crystal faces in minerals growing unconfined in silicate, aqueous or other fluids has been of interest to mineralogists since the nineteenth century. The monumental work of Dana *et al.* (1944) compiled the faces developed in a wide variety of minerals, with emphasis on the non-silicates. A more recent and comprehensive survey is that of Kostov and Kostov (1999). They give much more detail and applications of habit to mineral exploration and other aspects of geology. However, apart from plagioclase and diamond there are few quantitative studies of crystal shape.

4.4.1.1 Plagioclase

Plagioclase is the most abundant mineral in the crust; hence there is much interest in any petrographic parameter that can reveal more of its petrogenesis. In addition, plagioclase crystal size distribution has been studied in both volcanic and plutonic rocks and crystal shape is commonly determined in such projects (Figure 4.14). At first glance, controls on crystals shapes seem to be rather different in the microlites of volcanic rocks, as opposed to volcanic phenocrysts and plutonic crystals.

Crystallisation of microlites in volcanic rocks can increase gas pressure and initiate eruptions (Sparks, 1997, Cashman & Blundy, 2000) Hence, there has been much quantitative work on microlites. Experimental evidence suggests that there is a progression from tabular crystals at low undercooling ($\sim$30 °C) through acicular habits and finally to skeletal crystals (Lofgren, 1974). The

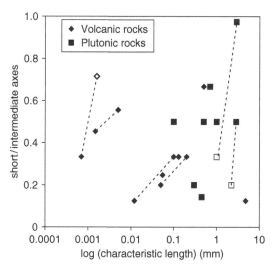

Figure 4.14 Short/intermediate axial ratios of plagioclase against characteristic or typical crystal length in volcanic and plutonic rocks (compilation of data cited here and in Section 3.4). Points from the same unit are linked. All crystals are close to tabular in form except for the open diamond, which is prismatic. The open squares are for foliated plutonic rocks.

shape of natural plagioclase microlites has been recorded frequently and many authors have noted that the shape of plagioclase microlites varies with their size: Cashman (1992) noted changes from an aspect ratio of 1:8 in the 1980 Mt St Helens blast dacite (typical length 10 μm) to 1:5 and finally 1:3 in the later erupted rocks (typical length 100 μm). Examination of her thin section photographs shows that the crystals are tabular. Higgins (1996b) found that crystals smaller than 200 μm in outline length have an aspect ratio of 1:5:6 whereas phenocrysts had a ratio of 1:3:4 (Figure 4.14). In both these cases the crystals are tabular: twin-plane orientations, where visible, indicate that the tabular face is parallel to {010}. Hammer *et al.* (1999) observed that the smallest microlites in a dacite are tabular (0.7 μm; 1:3:3), whereas the larger microlites are prismatic (1.6 μm; 1:1:7). They ascribed this to a change from a nucleation dominated crystallisation regime to one dominated by growth. These microlites are much smaller than those measured in other studies, hence there may be more complexity at this scale which is not seen in the larger microlites.

Compilation of plagioclase shapes from numerous sources reveals a broad correlation between the characteristic size of plagioclase crystals in volcanic rocks and their shape, as expressed by the short/intermediate ratios, with the exception of the smallest microlites (Figure 4.14). However, not all rocks have the same trend; hence it is possible that both size and shape track another parameter and the most likely is undercooling. Initially, a high degree of undercooling is required to initiate nucleation. Under such conditions the growth rate of the tablet edge is much greater than that of the tablet side. As crystal growth occurs, release of latent heat reduces the undercooling and the ratio of growth rates is reduced, hence leading to more equant crystals.

Plagioclase in plutonic rocks, such as anorthosite, gabbro, diorite and granodiorite does not generally have euhedral faces; however it commonly has a distinctive habit that depends on rock type and conditions of crystallisation (e.g. Higgins, 1991). Higgins (1991) examined plagioclase crystals from both massive and well-foliated anorthosites from the Sept Iles mafic intrusion and noted a significant difference in plagioclase shape. In the massive anorthosite the plagioclase had an overall axial ratio of 1:2:4, whereas in the laminated anorthosite the plagioclase was considerably more tabular with an axial ratio of 1:5:8 (the shapes shown here were recalculated from the original data). This suggests that growth under conditions of magma shear promotes the growth rate of the tablet edges (Higgins, 1991). In another study of plagioclase in anorthosite the crystal shape changed with coarsening from tabular crystals to more equant forms (Higgins, 1998).

It is possible that the crystal shape in both volcanic and plutonic rocks just reflects the chemical potential gradient around the growing crystal: where

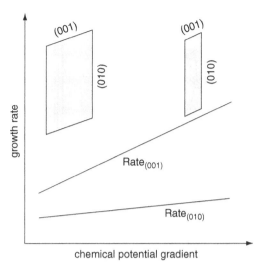

Figure 4.15 Possible response of plagioclase face growth rates to changes in the chemical potential gradient around the growing crystal (this study). The (001) face is taken to be typical of the tablet edge faces. The chemical potential gradient will be high at high degrees of undercooling as diffusion will be unable to supply nutrients to the crystal fast enough. The chemical potential gradient will be lower at lower undercooling where the crystals are growing more slowly or where shear advection removes depleted magma from around the crystal faces.

there is a high gradient the tablet edges are favoured, but under more equilibrium conditions the effect is reduced and growth is more equal on the tablet sides and faces (Figure 4.15).

4.4.1.2 Zircon

Zircon is a widespread, chemically resistant mineral that is used for geochronology and stable isotope studies (Hanchar & Hoskin, 2003). Zircon is not abundant in most rocks; hence crystals are generally extracted mechanically before analysis. The problem of extracted crystals is that they lose their petrographic context. However, zircons are tough and many retain their habit after extraction. Qualitative and quantitative studies of zircon habit has therefore received much attention (Figure 4.16; Pupin, 1980, Kostov & Kostov, 1999).

Zircons in rhyolite lavas and tuffs are prismatic and their length/breadth ratio varies from 1.3–1.5 for crystals 50 μm long to 2.1–3.0 for crystals 250 μm long (Bindeman & Valley, 2001, Bindeman, 2003). This may be an intrinsic property of zircons (Larsen & Poldervaart, 1957). Vavra (1993) has examined quantitatively the growth of individual zircons using both their overall shape

Zircon crystal forms

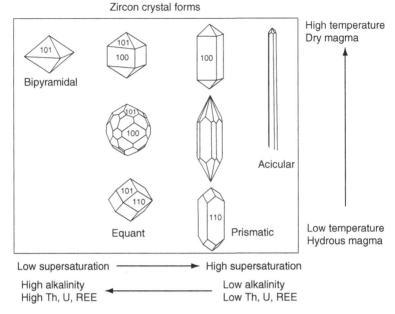

Figure 4.16   Controls on zircon morphology (Kostov & Kostov, 1999).

and internal zoning patterns. He again finds that the pyramidal faces in general grow faster than the prismatic faces as the undercooling (crystallisation driving force) increases, hence producing the same change in shape with crystal size that was observed qualitatively by other authors.

4.4.1.3 Other minerals

Euhedral quartz crystals commonly grow in hydrothermal metamorphic environments as prismatic crystals with pyramidal terminations. Ihinger and Zink (2000) used infrared spectroscopy to show how such crystals develop. They found that as the crystal grew the termination evolved from one dominated by $\{11\bar{1}1\}$ faces to a final form with $\{10\bar{1}1\}$ faces. They showed that an abundance of impurity phases such as HOH and AlOH can be used to measure the relative growth rates of the faces. However, it is not clear if the change in the relative growth rates of the faces was related to an increase in size or changing crystallisation conditions, such as temperature.

Crystals with rapid growth textures are found in many rocks and the most commonly described is olivine (Faure *et al.*, 2003). Up to ten different types of olivine morphology have been described (Donaldson, 1976) and these are broadly related to undercooling (supersaturation). The experiments of Faure *et al.* (2003) have shown that there are only three true morphologies and other observed shapes are just special sections. With increasing undercooling olivine

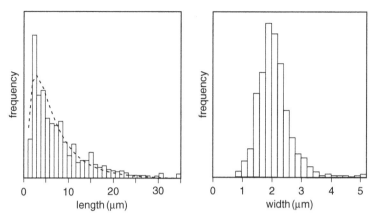

Figure 4.17   Shape of pyroxene microlites in rhyolitic obsidian (Castro *et al.*, 2003). The length of the crystals is much more variable then their width.

morphology went from tablets to hopper (skeletal) to swallowtail (dendritic). The development of the different forms is a function of both cooling rate and degree of undercooling.

Castro *et al.* (2003) have examined pyroxene microlites in rhyolitic obsidian using an innovative optical sectioning technique (see Section 2.2.2). They find that the crystal lengths vary over a factor of ten (3–30 μm), but the widths only vary by a factor of two (Figure 4.17). These pyroxene crystals initially develop as stubby prisms, with a width of 1–2 μm, and then just grow along the prismatic axis. Hence, the aspect ratio is proportional to the length. This behaviour can be produced if the ratio of the growth rates of prismatic faces over pyramidal faces decreases with undercooling.

Zeh (2004) investigated the shape of apatite grains in a series of meta-morphic rocks. He found that the ratio of the length/diameter (l/d) of the crystals and mean crystal length varied systematically with temperature of metamorphism. Those crystals from the lowest grade rocks had a mean l/d of 20 and a length of 1 μm and those from the highest grade had a mean l/d of 2 and a length of 100 μm. Allanite showed a similar change in shape, but it was much less pronounced.

Pallasites are unusual meteorites that are mostly comprised of olivine crystals set in a matrix of Fe–Ni alloys. In some pallasites the olivine crystals are euhedral or angular fragments, whereas in others they are well rounded. Saiki *et al.* (2003) showed experimentally that rounding of olivine could occur in geologically reasonable periods of time at a temperature of 1400 °C, which is sub-solidus for these compositions. Adjustments to the shape of the olivine grains occurred by volume diffusion in the crystal lattice and not by surface diffusion along the grain boundaries.

4.4.2 Deformed rocks

4.4.2.1 Sutured quartz grain boundaries

Sutured grain boundaries have been observed in many deformed rocks, but seem to be particularly evident on quartz–quartz boundaries (Figure 4.18). Where examined in thin section such grain boundaries are scale invariant over a length scale of one to two orders of magnitude (Kruhl & Nega, 1996). The fractal dimension varies from 1.05 to 1.30, but it should be remembered that this is the value for an intersection of the grain boundary with the plane of the thin section. The 3-D fractal dimension is not known, but should be related to these values.

During deformation at high temperatures there is equilibrium between the production of additional undulations of the grain surface and their removal by coarsening (ripening) processes as the texture approaches equilibrium. Hence the degree of suturing can be related to the strain rate of the rocks and their temperature. Kruhl and Nega (1996) analysed three disparate groups of rocks and proposed that the fractal dimension (D) can be used as a geothermometer: At a temperature of 350 °C $D = 1.28$ and at 700 °C $D = 1.08$. However, the strain rate is also a controlling factor (Takahashi & Nagahama, 2000). In addition, overprinting by subsequent events can produce scaling heterogeneities (Suteanu & Kruhl, 2002). Clearly, the application of this technique as a geothermometer, although useful, requires careful observation of other parameters.

Detailed examination of sutured quartz–quartz boundaries using a universal stage has shown that they are curved, but are composed of multiple straight segments (Kruhl & Peternell, 2002). The orientation of these crystal faces is

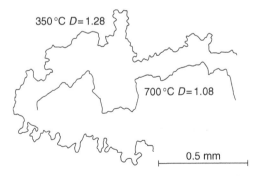

350 °C $D = 1.28$

700 °C $D = 1.08$

0.5 mm

Figure 4.18 Tracing of two sutured quartz–quartz grain boundaries (Kruhl & Nega, 1996). The outer curve was developed at 350 °C and has a fractal dimension of 1.28, whereas the inner curve developed at 700 °C and has a fractal dimension of 1.08.

controlled by the lattice orientation of the crystal and its neighbour. At low temperatures (~350 °C) the faces conform to a wide variety of indices, but as the temperature increases to 700 °C faces close to $\{10\bar{1}1\}$ are favoured. The presence of the crystallographically controlled faces probably makes the texture more resistant to addition changes, until the rock is deformed again.

4.4.2.2 Calcite in mylonites

Amongst the vast number of articles on the orientation of crystals in deformed rocks, there are only a few in which the shape of the crystals have been determined. Clearly, this can give information on deformation, like the size and orientation of the crystals. Herwegh *et al.* (2005) examined a series of carbonate mylonites from the Alps that formed at temperatures from 250 to 380 °C. They found that crystals from the low-temperature mylonites had a ratio of S/I of 0.8, which decreased to 0.45 for the highest temperatures. They also used the Paris shape factor to determine the sinuosity of the crystal outlines and found that it, too, increased with temperature. This seems to be the inverse of what was found in quartz mylonites (see above).

4.4.3 Dihedral angles

Dihedral angles can be used to explore the flow transport properties of fluid-bearing materials, hence there has been considerable experimental work in this field, but this has not been balanced by study of natural materials. In addition, there is a problem concerning the measurement methods: many early studies used apparent dihedral angles measured in sections that are difficult to assess in the light of the stereological studies of Elliott *et al.* (1997). Although 2-D studies form the bulk of the published work, such methods should only be used where 3-D methods cannot be applied, such as opaque minerals and small samples. Finally, there is some ambiguity in the way that dihedral angles are reported for polyphase rocks. Some authors record DAs as plagioclase–plagioclase–clinopyroxene: it is clear here that the DA of the last phase, clinopyroxene, is measured. However, others use terms like melt-solid-solid, in which the DA of the first term, the melt, is measured. There is little ambiguity for two phases, but for three phases the phase in which the DA is measured must be specified. I use here the convention A(B-C) to indicate that the DA was measured in A at a contact with B and C.

The earliest studies of dihedral angles in natural rocks were in granulites (Kretz, 1966b, Vernon, 1968, Vernon, 1970). Clinopyroxene (olivine–olivine) and clinopyroxene (plagioclase–plagioclase) intersections both have median dihedral angles of $115° \pm 17°$ (1 sigma). This indicates that there is little contrast between the interfacial energies of these three minerals under these conditions.

Many earlier DA studies were centred around three themes: basaltic melts in peridotites, partial melting of crustal silicate rocks and H_2O–CO_2 crustal fluids. Commonly, these studies were experimental with relatively few studies of actual rocks. The distribution of melt in peridotite is of interest for the origin of basalt by partial melting (e.g. Faul, 1997) in addition to the geophysical properties of the mantle, such as seismic attenuation and mechanical strength. The DAs of basalt (olivine) and other minerals have been investigated in both experimental and natural materials (e.g. Waff & Bulau, 1979, Cmiral *et al.*, 1998, Duyster & Stockhert, 2001). In general, olivine crystals surrounding melt pockets are faceted, not curved, but the dihedral angles are low to zero (Cmiral *et al.*, 1998). Other minerals in mantle rocks have a significant effect on the overall permeability: diopside and spinel increase permeability whereas enstatite lowers it (Schafer & Foley, 2002). Hence, melts should flow more easily through spinel lherzolites than harzburgites.

Many types of granite originate by partial fusion of the crust, followed by aggregation and escape of the magma. Laporte and Watson (1995) have examined the process using experimental and theoretical techniques, with special emphasis on the role of biotite and amphibole. They concluded that crustal rocks cannot be modelled as ideal systems because they are comprised of many different minerals, mostly with medium to high degrees of anisotropy of interfacial energy. The overall fabric (crystallographic preferred orientations) and heterogeneity of the rock also had an important effect. For rocks with highly anisotropic minerals, such as biotite, and low degrees of melting, the silicate liquid lies in isolated plane-faced pocket at grain junctions. The shape of these pockets is controlled by the anisotropy of the minerals and the texture of the rock. The low DA for liquid (solid–solid) junctions in crustal rocks indicates that melts should be connected at very small degrees of melting and hence should be continuously drained from a rock during melting. Amphibolites may have a connectivity threshold of 3%. However, such melts are very viscous and hence viscosity and not connectivity may limit the separation of melt and restite. In contrast, well-foliated biotite rich rocks may act as barriers to transverse liquid flow.

The role of fluids, particularly those rich in water, in geological processes is pivotal. Holness (1993) examined experimentally the behaviour of H_2O–CO_2 fluids in quartzites. Water-rich fluids have DAs close to or lower than the critical value of 60° at both low and high temperatures, which was attributed to the presence of a film of water on the quartz surface. This implies that there is a window of permeability at both low and high temperatures for infiltration of water. The behaviour of CO_2 was completely different: it had high DAs at all temperatures suggesting that it was not absorbed onto the quartz. The presence of solutes like NaCl appears to reduce further the DA of aqueous solutions and hence enlarge the window of permeability (Watson & Brenan, 1987).

Although processes during the early stages of accumulation of crystals in magma chambers are reasonably well understood, there is little information on the later textural evolution of these rocks. On the basis of crystal size distributions Higgins (2002b) and Boorman *et al.* (2004) both considered that textural evidence for crystal accumulation is frequently overprinted by coarsening. Holness *et al.* (2005) have attacked the problem by the analysis of DAs. They pointed out that initial growth should generate impingement textures with a median dihedral angle of about 60° (Figure 4.5). Analysis of melt (clinopyroxene–clinopyroxene), melt (amphibole–amphibole) or melt (plagioclase–plagioclase) DAs in glassy lavas established a range of median DAs from the impingement values down to 28°. For the Kula lavas DAs and changes in crystal shape indicated that textural equilibration was achieved. However, for the Kameni lava they favoured diffusion limited growth. Cumulate plutonic rocks do not contain any melt, but some minerals may pseudomorph former melt pockets. Hence, the next step was to examine the DAs of such minerals in cumulate rocks.

The small layered intrusion on Rum, Scotland has received enormous attention, but there are still major uncertainties in its petrogenesis: one such problem is the origin of peridotite layers. These may have formed as accumulations of olivine or by injections of olivine-rich picrite into the semi-solid layered gabbros and troctolites. Holness (2005) has investigated the textures of the rocks that host the peridotite layers to determine their thermal history and hence resolve their origin. She considered that clinopyroxene pseudomorphs the last melt and measured the clinopyroxene (plagioclase–plagioclase) (CPP) DAs along a section orthogonal to a peridotite layer. She found that all CPP median DAs lay between the values for impingement textures (~60°) and the values for complete equilibrium (115–120°, Figure 4.19). It is assumed that higher median values reflect movement towards solid-state equilibration during reheating by the injection of peridotite sills. In most of the traverses there was no correlation between median DAs and proximity to peridotite layers, except for one layer. Hence, both models for the origin of the peridotite layers are validated, but for different layers.

Sulphide minerals commonly occur as globules in silicate rocks. Mungall and Su (2005) showed experimentally that this was because of the high surface energy between sulphide and silicate liquids, as compared to that between silicate minerals and silicate liquids. They considered that at low degrees of

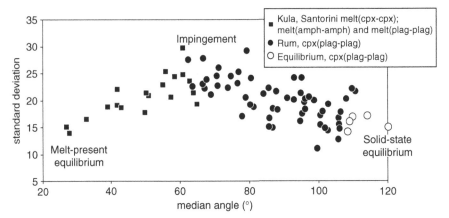

Figure 4.19   Median against standard deviation of dihedral angles in melt-present and solid materials (Holness *et al.*, 2005). The Kula and Kameni data (squares) are from partially melted crystalline enclaves. The data from Rum, Scotland (filled circles) are from solid gabbro and troctolite (Holness, 2005). Granulite samples (empty circles) provided data for solid-state equilibrium (Kretz, 1966b, Vernon, 1968, Vernon, 1970).

supersaturation sulphide droplets may nucleate at wide intervals leading to kinetic control of composition. The high surface energy of sulphide droplets will hinder their migration through a crystal mush, unless propelled by moving silicate liquids. Clearly, this has important implications for the origin of massive sulphide bodies.

Notes

1. The terms crystal habit, crystal shape and crystal morphology are considered to be synonyms.

5

Grain orientations: rock fabric

5.1 Introduction

The quantitative study of grain orientations (fabric)[1] is very important in structural geology and materials science (Wenk, 1985, Kocks *et al.*, 2000, Randle & Engler, 2000, Wenk, 2002, Wenk & Van Houtte, 2004). It is less established in igneous petrology (Nicolas, 1992, Smith, 2002), where fabric has been used mainly to establish flow directions and mechanisms in lava flows and dykes. Fabric studies find their way in geophysics as there is considerable interest in the anisotropy of seismic velocities (e.g. Mainprice & Nicolas, 1989, Xie *et al.*, 2003). Fabric is also important in engineering geology, as it affects the physical properties of rocks (e.g. Akesson *et al.*, 2003). It is also of interest in palaeontology and medicine for quantifying the structure of bone (Ketcham & Ryan, 2004). Knowledge of grain orientations is also necessary to calculate size distributions from the dimensions of grain intersections in all the domains already mentioned (see Section 3.3.4).

The grains that define the fabric of a material can be a variety of different objects: commonly crystals, clasts and enclaves. One measure of fabric is based on the orientations of the grain shapes (shape preferred orientations: SPO). Here the nature of the material is unimportant for the measurement of the fabric. However, the contrast in viscosity between the grain and matrix during production of the fabric will define what parameter is recorded by the fabric. For example, in magma originally spherical deformable enclaves will record strain, whereas rigid crystals will rotate and just record indirectly aspects of the deformation.

If the grains studied are crystals then another measure is possible: the fabric can be defined by the orientation of the crystal lattices, here termed lattice preferred orientations or LPO (also known as crystallographic preferred orientations; macrotexture or texture in materials science literature). Of course, if crystals are euhedral then these two orientations are parallel. This is generally not the case in most metamorphic and many plutonic rocks, as crystals do not grow in an unobstructed environment or they may have been partly dissolved. In this situation orientations defined by crystal shape and lattice will not necessarily be related.

5.2 Brief review of theory

5.2.1 Fabrics developed during growth of new crystals

5.2.1.1 Crystallisation from a magma

Isolated crystals in a homogeneous liquid nucleate with random orientations. If there is no flow of the magma, then growth will conserve the orientation of the nuclei and there will be no development of fabric. However, if the growth of crystals is constrained by the presence of other crystals, or if there is a significant concentration gradient in the magma then a fabric may develop during crystallisation. Such fabrics are growth phenomena, hence they are both shape and lattice preferred orientation fabrics.

The most common type of crystallisation fabric in plutonic rocks is comb-layering (Lofgren & Donaldson, 1975). This texture is produced where crystals initially nucleate on a rigid or semi-rigid plane. The orientation of the nuclei is not important. If the crystals have an initial shape anisotropy then competition between the crystals will eliminate those with maximum growth rates parallel to the crystallisation plane. This leads to a strong linear fabric, normal to any structures that define the original solidification front, such as mineral layering. Crystals may become curved or branching if they can grow fast enough to escape the zone depleted in nutrients close to the solidification front. Similar fabrics can be developed in some volcanic rocks: the most spectacular are the spinifex textures in komaiites (e.g. Shore & Fowler, 1999).

Equilibration of magmas or rocks is the process by which the surface and internal energy of the system is minimised. One process is by the coalescence of mineral grains, which reduces the total surface energy. Energy is further minimised if the differences between the orientations of the crystals is reduced by rotation of grains. There is evidence for this in the commonly observed phenomena of synneusis of igneous minerals such as olivine and plagioclase (see Section 6.2.3; e.g. Schwindinger & Anderson, 1989). Recently evidence has been found that some garnet porphyroblasts are also formed by coalescence and rotation of many grains (Prior *et al.*, 2002).

5.2.1.2 Solid-state mineral crystallisation

When crystals nucleate and grow in a solid medium then the orientation of the crystal will be organised so as to minimise the energy required for growth. Hence, growth of crystals (or transformation of existing minerals) will generate fabrics that are defined by both the crystal lattice and shape. If deformation is coaxial then SPO and LPO will be parallel. Under non-coaxial deformation the obliquity between the LPO and SPO can be used to determine the sense of shear. Unless the deformation that lent the crystals their orientation ceases immediately afterwards the newly formed crystals will be deformed in the solid state as they grow.

5.2.2 Fabrics produced during magmatic and solid-state deformation of existing crystals

5.2.2.1 Property changes during solidification

The solidification of magma and the melting of rock, are not simple processes, but involve a transition in properties from a liquid to a solid. Three stages are usually recognised but the transition points appear to be different for solidification and melting (e.g. Vigneresse *et al.*, 1996, Arbaret *et al.*, 2000). During solidification from 0 to 30%–40% crystals the magma behaves as a suspension, with pure Newtonian behaviour, especially at low crystal concentrations. Above 30%–40% crystals interact frequently and the magma acquires yield strength. The magma can now be considered to behave as a Bingham fluid. A rigid framework of crystal chains, or more accurately foam-like crystal surfaces, may develop in some rocks at this stage, or even earlier (Philpotts *et al.*, 1999). Otherwise, the quantity of crystals must increase to 50%–65% crystals before a continuous network of crystals forms and the yield strength increases drastically. The magma then has a plastic behaviour that is close to that of a solid. At this point strain may be concentrated into discrete zones, crystals may fracture or be divided into sub-grains and folding may occur. Clearly the first stage must be considered to be in the igneous domain. The last stage is more metamorphic, but the intermediate stage is difficult to classify. In many plutonic rocks deformation continues through all these stages maintaining the same orientation.

During melting the rock becomes much weaker with as little as 5% melt (95% crystals) as melt pockets become interconnected (Rosenberg & Handy, 2005). A second major rheological change occurs about 30% melt after which the material resembles magma. The differences between solidification and melting probably reflect textural changes: for instance during melting small crystals will be removed first, leaving isolated larger crystals. Whereas during

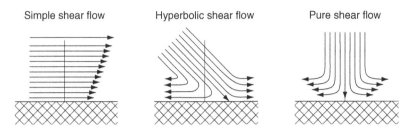

Figure 5.1   Flow lines for different shear flow types. Simple and hyperbolic shear are non-coaxial shear; pure shear, or compaction, is coaxial shear.

solidification small crystal will nucleate and grow as long as significant under-cooling is maintained.

Deformation can occur in both rocks and magmas. Plastic and solid-state deformation of crystals not only changes the shape of crystals, but also their size. The basic theory of these processes is discussed in Section 3.2.5. Magmatic deformation will only change the orientation and position of crystals, but it depends on the type of shear regime. Shear can be classified with reference to a fixed plane (Figure 5.1). Simple shear is parallel to the plane, pure shear is normal to it and hyperbolic, or mixed, is between the two extremes.

5.2.2.2 Magmatic deformation

Numerical models of magmatic deformation are based on the motions of rigid particles in a fluid. There are a number of assumptions (Arbaret *et al.*, 2000, Iezzi & Ventura, 2002). (1) Particles are ellipsoids or parallelepipeds of varying dimensions. (2) They do not interact with each other. (3) They do not change shape during deformation. (4) The liquid is Newtonian and flows constantly without turbulence. The most recent numerical modelling was done in three dimensions using particles of different shapes and shear regimes (Iezzi & Ventura, 2002). Under simple and hyperbolic shear, cubes and elongated tablets never achieved a stable configuration: they just continued to roll (Figure 5.2a). Simple 2-D analogue modelling shows that at high concentrations of particles interactions between adjacent grains prevents cyclical movements and grains become imbricated (Ildefonse *et al.*, 1992). Tablets and prisms rotated towards the plane of shear and produced a strong fabric (Iezzi & Ventura, 2002). For pure shear (compaction) all shapes rotated until they were parallel to the shear plane (Figure 5.2b). However, this process is not very efficient and high degrees of deformation are necessary to produce a visible fabric (Higgins, 1991, Ildefonse *et al.*, 1992).

The lack of stability of fabrics for grains with equant or near-equant shapes means that local fabrics may not indicate the direction of shear. Iezzi and

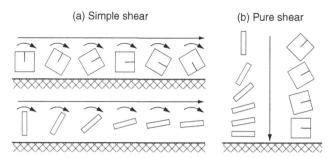

Figure 5.2   Schematic behaviour of a cube and tablet in a sheared magma.

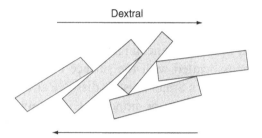

Figure 5.3   Imbrication or tiling of grains or crystals for dextral simple shear. Rotation of individual crystals in response to simple shear is stopped by the presence of an adjoining crystal.

Ventura (2002) showed that the strain regime can only be determined precisely if the orientations of crystals with different shapes can be determined. Normally this would be for different minerals.

Mineral grains do usually interact, especially at higher grain concentrations, but this has not been modelled numerically yet. One result of such interactions is tiling or imbrication (Figure 5.3; e.g. Blumenfeld & Bouchez, 1988). This is when the rotation of one grain is stopped by impact against another grain. Two-dimensional analogue models indicate that tiling can be used to identify shear directions and planes (Ildefonse *et al.*, 1992). Nicolas and Ildefonse, (1996) have proposed that crystal mushes in magma chambers flow and develop fabrics after deposition without plastic deformation. They suggest that magmatic flow of crystal-rich material (>80% crystals) is enabled by pressure-solution at imbrication points: flow will be stopped until sufficient material has been removed and the crystals are no longer imbricated. Flow then resumes until imbrication is again encountered.

Magmatic deformation is controlled by the shape of the grains; hence the fabric is the result of SPO. If the crystals are the results of growth without significant solution, then their shape will reflect the orientation of the crystal lattice and the LPO will be parallel to the SPO.

Many magmatic rocks contain enclaves – generally rounded patches of texturally or compositionally contrasting rocks. Enclaves are generally assumed to result from the mixing of two or more magmas. If the enclaves have a significantly higher liquidus temperature than the magma then the enclaves will be chilled and act as rigid bodies during magmatic deformation, similar to the crystals discussed above. However, if the compositional contrast is small then there will be little viscosity contrast between the enclaves and the magma. If the enclaves were originally spherical then their shape will record the total magmatic strain.

5.2.3 *Influence of coarsening and pressure solution compaction on fabric*

Coarsening (Ostwald ripening, annealing, textural maturation) is the process by which the energy associated with the surface of the crystals in a rock is minimised (see Section 3.2.4): Small crystals dissolve at the same time as large crystals grow. This process may be important in many igneous rocks and metamorphic rocks that are not being deformed.

As the large crystals grow they need to be accommodated within a fixed volume of rock. They will tend to rotate to fit into the space created by the solution of the smaller crystals. If the initial rock has a strong shape-preferred foliation then this process will reduce the quality of that foliation (Higgins, 1998).

Pure shear, or compaction, is not a very efficient process at producing a significant fabric (see the above section). An alternative is pressure-solution compaction (Meurer & Boudreau, 1998). Here, the movement of the crystals under pure shear is augmented by recrystallisation: plagioclase grains that are parallel to the applied stress will dissolve more readily than those that are normal to the stress. This is the same process that is seen in the diagenesis of sedimentary rocks and occurs because of the excess energy where two grains are in contact. This process will yield a SPO, but not an LPO. However, an existing LPO could be preserved. Nicolas and Ildefonse (1996) have proposed that pressure-solution may also augment simple shear of crystal-rich magmas (see Section 5.2.2.2)

5.3 Introduction to fabric methodology

A large number of analytical methods have been used to examine the orientation of grains in rocks (Table 5.1). Some methods measure the orientation of the grain shape (SPO), others quantify the orientation of the crystal lattices (LPO) and other methods look at a combination of shape and lattice orientation. Some methods examine the orientation of each individual crystal,

Table 5.1   *Summary of grain orientation analytical methods.*

Analytical method	SPO or LPO	Individual or bulk data	Volume or section
Serial sectioning	SPO	Individual	Volume
X-ray tomography	SPO	Individual	Volume
Sections: optical, orthogonal stage	SPO or partial LPO	Individual	Section
Sections: universal stage	LPO	Individual	Volume
Electron backscatter diffraction	LPO	Individual	Volume
Anisotropy of magnetic susceptibility (AMS)	LPO and SPO	Bulk	Volume
Anisotropy of (partial) anhysteretic remanent magnetism (AARM – ApARM)	LPO and SPO	Bulk	Volume
Neutron diffraction	LPO and SPO	Bulk	Volume
X-ray diffraction	LPO and SPO	Bulk	Volume

whereas others only give a measure of the overall or bulk grain orientations. Grain orientation is a three-dimensional (volume) property but orthogonal examination of thin sections only gives orientations in the section; hence several nonparallel sections must be examined.

5.4 Determination of shape preferred orientations

5.4.1 Direct, three-dimensional analytical methods

The orientation of individual grains can be determined by serial sectioning (see Section 2.2.1) or X-ray tomography (see Section 2.2.3). Voxel or vector images can be analysed using the methods mentioned in Section 2.3: each grain is approximated by a triaxial ellipsoid and these ellipsoids summed to give the overall fabric ellipsoid. So far, there are no orientation studies yet published using these techniques. In transparent materials, such as volcanic glasses, the orientation of crystals can be measured directly with a regular or confocal microscope (see Section 2.2.1; Manga, 1998).

5.4.2 Section and surface analytical methods

Orientation data can be easily acquired from rock surfaces, slabs and sections, using a variety of techniques described in Section 2.2. Such images can be processed manually or automatically to produce classified mineral images (see Section 2.3). The classified images can be processed to extract overall orientation data using a number of methods described below. Data from orthogonal sections can be combined to give 3-D orientations.

Imbrication has been quantified by simply counting the imbrication directions of suitable pairs of touching grains (e.g. Ventura *et al.*, 1996). A significant difference between the numbers of pairs that indicate dextral (clockwise) and sinistral (anticlockwise) movements is used to define the overall sense of simple shear. If the flow direction can be established by other means (e.g. geological structures or palaeosurface orientation) then the grain orientation can be used to determine the imbrication orientation and hence the direction of flow.

5.4.3 Extraction of orientation parameters

Orientation analytical data need to be summarised before they can be applied to geological problems. Most work in this subject has been applied to sections, but can be extended easily to three-dimensional data. A number of different approaches are possible: for instance individual crystal outlines can be summarised by the inertia tensor method, whereas the intercept method looks at the orientation distribution of the grain outlines.

5.4.3.1 Inertia tensor method

Any complex grain outline can be reduced to an equivalent ellipse that has the same area and moment of inertia (Figure 5.4). The orientation ellipse of each separated grain in a section is calculated by the inertia tensor method and the results summed for all grains. Clearly, the method can only be used where the individual grains have been separately defined. The method can also be extended to 3-D structures. The inertia tensor is calculated from the coordinates of the grain outlines, or from the pixels of a bitmap image (Launeau & Cruden, 1998). The inertia tensor method has the advantage that the contribution of each grain to the total orientation ellipse is weighted by its area (see Appendix; Launeau & Cruden, 1998).

5.4.3.2 Intercept (intersection) method

The intercept method is a powerful technique for the numerical analysis of fabrics (see Appendix; Launeau *et al.*, 1990, Launeau & Robin, 1996). It actually quantifies the orientation of the boundary of a phase and hence does not necessarily give the same information as the inertia tensor method. Because it only examines boundaries, it can be used to analyse any set of objects (crystals, clasts, voids, lineaments, etc.) that can be identified in the image. It is not

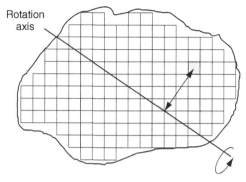

Rotation
axis

Figure 5.4 The inertia tensor parameter of a crystal outline. The method is equivalent to the following process: the crystal outline is considered as a thin sheet of material. The moment of inertia of the crystal outline for an axis in the plane of the paper can be calculated by summing the inertia contributions of the pixels enclosed by the outline. For each pixel the inertia depends on the distance from the rotation axis. The moment of inertia is calculated for different orientations of the axis and the maximum and minimum values recorded. These are used to calculate the parameters of an equivalent ellipse.

necessary that the objects are all separated, as is the case for the inertia tensor method described above. For example, if a slab of granite stained for potassium feldspar with sodium cobaltinitrite is scanned, then quartz, potassium feldspar and plagioclase can be readily distinguished from each other. However, most of the larger crystals will touch, giving a framework. The intercept method can be used to determine the SPO of this texture. It should be noted that the intercept method is more commonly used in stereology to count objects or determine the ratio of surface area to volume (see Section 2.5.2; Russ, 1986).

The method is based on counting intercepts (intersections): a test line is drawn through the section and an intercept is counted where the object in question is exited (Figure 5.5a). In practice many test lines are drawn through the section at regular intervals and at different angles. The numbers of intercepts are counted for each test line and summed for each angle. The intercept totals for each angle are divided by the total length of the test lines, giving the intercept density. The inverse of the intercept density is the mean intercept length. Both data sets can be plotted as rose diagrams. The orientation or strain ellipse can be derived from the rose of mean intercept lengths.

The star length and star volume methods are intrinsically similar to the intercept method (Smit *et al.*, 1998). A point is selected randomly within the phase of interest (Figure 5.5b). The lengths of lines at different angles passing through this point are compiled. The star length for an angle is the mean length of the lines at that orientation. The star volume method is similar, but the cube of the length is summed giving a different weighting for different crystal sizes.

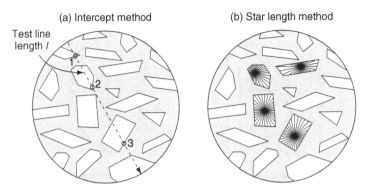

Figure 5.5   (a) The intercept method. A test line of length *l* is drawn and the number of times that a crystal is exited (intercepts) is counted, three in this case. (b) The star length method. A point is randomly placed within the phase of interest. The lengths of lines radiating from this point at particular angles are compiled.

The method has also been extended to the analysis of 3-D voxel images (Ketcham & Ryan, 2004).

5.4.3.3 Grey-scale image analysis methods

Rock fabrics in two dimensions can be deciphered at different scales using powerful wavelet analysis techniques (Darrozes *et al.*, 1997, Gaillot *et al.*, 1997, Gaillot *et al.*, 1999). This technique uses 2-D wavelets to give estimates of the orientation and quality of orientation of mineral grains at different scales. It is the only generalised method that can do this. It is not necessary to produce a classified mineral image as the method can be applied to grey-scale images. Although the method has much promise it has not been used much.

Another image analysis technique that can be applied to grey-scale images is based on the autocorrelation function (Heilbronner, 1992, Heilbronner, 2002). This function is used to identify periodic structures in images. It is less generalised and powerful than wavelet-based methods. However, it is better supported by software. A similar method is the directional variogram (Cressie, 1991). It is defined by the variance of the grey-scale pixels at a distance *h* in a direction *a*.

5.5 Determination of lattice preferred orientations

5.5.1 Optical analytical methods: regular and universal stages

Partial information on the lattice orientation of some minerals can be obtained from a regular thin section and petrographic microscope. For uniaxial minerals, such as quartz and calcite, the optical axes are parallel to the crystal-lographic axes; hence extinction positions in cross-polarised light can indicate the orientation of a plane that contains the extraordinary ray or rotational axis of symmetry.

For biaxial minerals the optical axes are not generally parallel to the crystal-lographic axes, hence structural features such as twins or cleavage are necessary to define the lattice orientation. Some information on the lattice orientation of minerals can be readily obtained from the orientation of the trace of the twin lamellae. This is most commonly applied to plagioclase.

The universal stage is an accessory for a petrographic microscope that enables a standard thin section to be rotated out of the plane of the microscope stage. A standard thin section is sandwiched between two glass hemispheres and the whole assemblage can then be rotated along four or five independent axes (Emmons, 1964, Muir, 1981). Most universal stages have to be read manually, but some have been modified by the addition of sensors that transfer data to a computer. Two or three orthogonal sections must be measured, as the section can only be rotated up to 45° out of the plane of the stage.

It is relatively easy and quick to measure the orientations of uniaxial minerals such as quartz and calcite, as only one direction need be determined. Orthorhombic minerals, such as olivine, are more complex as three directions must be determined. The most difficult are monoclinic and triclinic minerals, such as plagioclase. The relationship between the lattice orientations and the axes of the optical indicatrix is controlled by the composition of the plagio-clase. If the composition is known then the lattice orientation can be determined from measurements of optical axes, cleavage planes and twin composition planes. The method is complex but a computer program can ease the calculations (Benn & Mainprice, 1989). Normally grains are measured in three orthogonal sections. However, Ji *et al.* (1994) suggest that measuring all twinned grains in any two orthogonal sections is more statistically meaningful and time-saving. However, their method does not always provide a complete LPO.

5.5.2 *Computer-integrated polarisation microscopy*

Partial information on the orientation of anisotropic grains in a section can be obtained from their extinction positions in cross-polarised light. A computer-controlled microscope stage has been developed that can automate this process (see Section 2.6.4; Heilbronner & Pauli, 1993, Fueten, 1997, Fueten & Goodchild, 2001). This system can give information on individual grains as well as the overall fabric ellipsoid of the rock. In general, it can only be applied to rocks containing a single mineral with an orthorhombic or higher symmetry. Plagioclase, for example, cannot currently be measured using this technique, because of its low symmetry and almost ubiquitous twinning.

5.5.3 *Electron backscatter diffraction (EBSD)*

Electron backscatter diffraction (EBSD) is a versatile technique that can give the complete crystallographic orientation of individual mineral grains as small as 1 µm in polished thin sections and surfaces (see review in Prior *et al.*, 1999). It can be used with all minerals, even those with low symmetry, such as plagioclase. It uses a standard scanning electron microscope (SEM) equipped with a phosphor screen and a light detector. Many systems are completely automated and can collect data from crystals in an area of 5 mm^2.

When an electron beam strikes a surface, electrons will be scattered by the atoms of the material in the top few µm (see Section 2.5.2.1). This omnidirectional population of electrons is the source for EBSD in an SEM. Electrons are diffracted by lattice planes in a crystal to produce two cones, with axes normal to the lattice plane. Under normal conditions for EBSD the opening angle between the two cones is very small; hence the electrons appear to stream out along the lattice planes. These electrons are allowed to strike a phosphor screen where they are converted into light. This light is detected by a CCD and conveyed to the computer. Many lattice planes in a crystal will diffract electrons and these bands will interfere constructively. Where a number of bands intersect, a bright spot will form which corresponds to a zone axis. Crystallographic directions can be identified from the relative position of the bands and zones. This process is commonly automated, although there is always a proportion of patterns that cannot be indexed. Spatial resolution is about 1 µm in geological specimens, but can be improved by using a field emission electron gun.

There are always compromises to the optimal configuration for EBSD, especially as the necessary hardware is commonly added onto an existing SEM. The sample surface is oriented at an angle of about 70°. The phosphor screen must be as close as possible to the part of the sample illuminated by the electron beam. It is better to move the sample and not the beam so that the geometry remains constant.

The importance of correct sample preparation cannot be overstated. EBSD observes the crystal lattice in the top 1–2 µm of the surface; hence the lattice must be intact there. Crystal faces and fractures provide a pristine lattice, but are not very useful for most projects. Mechanical polishing generally damages the lattice, although some workers have had success with a very fine colloidal silica (Xie *et al.*, 2003). A number of other treatments can be used: etching removes the damaged material, but generates unwanted relief. Chemical-mechanical polishing gives a good surface, but some minerals dissolve rapidly in the fluid used (Prior *et al.*, 1999). Ion-beam milling is another promising possibility.

Another problem for geological specimens is that they are generally non-conducting and hence surface charging may occur. This effect is very variable both spatially and temporally. Coating the sample in carbon reduces the problem, but requires higher beam currents. Most studies do not use coated samples, but just accept that some measurements will be lost owing to charging. Another possibility is to maintain the sample chamber at a higher pressure in an environmental SEM (Habesch, 2000). This results in more electron scattering but less sample charging.

5.6 3-D bulk fabric methods: combined SPO and LPO

A number of methods examine the bulk physical properties of the grains of one or all phases in a rock. In general such methods record a combination of the SPO and LPO. They give a single value for the orientation ellipsoid for the material.

5.6.1 Anisotropy of magnetic susceptibility

Anisotropy of magnetic susceptibility (AMS) is a technique for measuring the bulk, 3-D magnetic fabric of rocks. It is sensitive, quick, and relatively cheap and can be applied to a wide range of rock types (Jackson, 1991, Rochette *et al.*, 1992, Tarling & Hrouda, 1993, Rochette *et al.*, 1999, Martin-Hernandez *et al.*, 2005). It has been applied extensively in all branches of structural geology and petrology. If the susceptibility is dominated by magnetite then the AMS measures only the SPO and spatial distribution of that mineral. If there are significant contributions from paramagnetic minerals then the AMS records a mixture of the SPO and LPO. AMS is measured using core samples 25 mm in diameter and 22 mm long. The cores are oriented in the field or drilled in the laboratory from oriented blocks. Many samples are generally taken at each outcrop or unit of interest and the results averaged. AMS is measured in a dedicated instrument.

The total magnetic susceptibility of a sample, K, is the ratio of the intensity of magnetisation of a material divided by the strength of the inducing magnetic field. It is not necessarily related to the remanent magnetism, which is the magnetic field of a sample in the absence of an inducing field. The magnetic susceptibility is commonly anisotropic and this anisotropy has an orientation: hence, AMS is second order tensor, K_{ij} and can be represented as an ellipsoid.

Three groups of minerals contribute to the susceptibility (Rochette *et al.*, 1992). (1) Ferrimagnetic minerals include those that are generally considered to be 'magnetic', that is they have a significant remanent magnetism. They have a much higher susceptibility than other minerals and will dominate the susceptibility of a rock when the measurement is carried out under normal conditions, in a low field. The most well-known ferrimagnetic minerals are magnetite, ilmenite, pyrrhotite, garnet and hematite. (2) Paramagnetic minerals have lower susceptibilities, but may be present in larger quantities than ferrimagnetic minerals. Those minerals that are richer in iron will generally have a higher susceptibility. Important minerals of this group are pyroxenes, amphiboles, staurolite, cordierite and biotite. (3) Diamagnetic minerals have a very low negative susceptibility; this means that the induced field has an opposite polarity to the applied field. These minerals are only important in the absence of ferrimagnetic or paramagnetic minerals. Prominent diamagnetic minerals are quartz and calcite. The total anisotropy of magnetic susceptibility of a rock is produced by the anisotropy of magnetic susceptibility of the mineral grains, if they have a LPO. It is also produced by the SPO of minerals which have a significant total susceptibility, but which are not necessarily anisotropic (e.g. garnet).

AMS analysis produces a susceptibility ellipsoid whose principal axes are labelled $k_1 \geq k_2 \geq k_3$ (or k_{max}, k_{int} and k_{min}). The parameters of the ellipsoid are commonly reduced to orientation, mean susceptibility ($k_m = (k_1 + k_2 + k_3)/3$), lineation parameter ($L = k_1/k_2$), foliation parameter ($F = k_2/k_3$) and degree of anisotropy ($P = k_1/k_3$). The shape of the magnetic anisotropy ellipsoid can be displayed on a 'Flinn-type' diagram (Flinn, 1962), where L is plotted against F.

Most interpretations of AMS data are based on the following assumptions (Borradaile, 1988, Rochette *et al.*, 1992, Borradaile & Henry, 1997). (1) The AMS ellipsoid is coaxial to the fabric, with the k_3 axis perpendicular to the foliation and the k_1 parallel to the lineation. (2) The shape of the AMS ellipsoid is related to the rock fabric, in terms of intensity of foliation and lineation development. (3) AMS measurements are not affected by remanent magnetism. These assumptions must be verified by comparison with independent mineral fabric measurements using any of the methods described in this chapter. Where they are correct the magnetic fabric is called 'normal'. In some cases it is found that the k_1 axis is perpendicular to the foliation and the k_2 axis is parallel to the lineation. This is called a 'reverse magnetic fabric'. In other cases k_2 is perpendicular to the foliation, which is referred to as an 'intermediate magnetic fabric'. Finally, the magnetic axes may have no relation to the structure.

In some situations reverse or intermediate magnetic fabrics may have been identified on structural, rather than mineral fabric criteria. Hence, it is possible that the magnetic fabrics are in fact normal and the complexity of the structure has been underestimated (Rochette *et al.*, 1992). Reverse and intermediate magnetic fabrics can be identified by rock and magnetic fabric measurements (Geoffroy *et al.*, 2002); however, in these cases the magnetic measurements are made redundant by the fabric analyses. A more useful way of detecting reverse and intermediate fabrics is by identification of the magnetic minerals that produce the susceptibility. A number of minerals, such as single-domain magnetite and tourmaline, can produce an inverse fabric (Rochette *et al.*, 1992). The magnetic mineralogy can be identified using optical means, but commonly many magnetic minerals are present. In these cases other magnetic methods can be used (see Section 5.6.2).

5.6.2 Anisotropy of (partial) anhysteretic remanent magnetisation

Many rocks have complex histories which are reflected in more than one fabric. In many situations a weak, late, geologically insignificant magnetic fabric can dominate over fabrics developed in response to more important geological events. The AMS of a sample is contributed by all iron-rich minerals. However, anisotropy of anhysteretic remanent magnetisation (AARM = ARM = anisotropy of isothermal remanent magnetism) and the related method anisotropy of partial anhysteretic remanent magnetisation (ApARM = pARM) respond only to ferrimagnetic minerals, essentially magnetite, hematite and pyrrhotite (Jackson *et al.*, 1988, Jackson, 1991).

Remanent magnetism is the magnetic field produced by a sample in the absence of an applied field. It is produced by the ferrimagnetic minerals and can be used in two different ways. The direction of the natural remanent magnetism of samples (NRM) is used in palaeomagnetic studies to establish the direction of the magnetic poles with respect to the sample. Similarly, if other data are known then the NRM can be used to correct for tilting and rotation of the unit since the fabric was produced (e.g. Palmer & MacDonald, 2002). The NRM of a sample can be erased by thermal or alternating field demagnetisation and a new magnetism induced. This magnetic field will be related to the anisotropy of the ferrimagnetic minerals in the rock. AARM examines the total anisotropy of remanence in a rock, produced by all ferrimagnetic grains. The strength of remanence is different for each grain and this can be used in ApARM to separate populations of grains and examine their fabric independently.

The strength of remanence is measured by the coercivity, which is the intensity of the magnetic field required to reduce the magnetisation of that material to zero after the magnetisation of the sample has reached saturation. 'Hard' materials have a high coercivity and 'soft' materials a low coercivity. The coercivity depends mostly on the nature of the mineral, its size and shape. The fabrics of grains with different coercivities can be determined by applying a weak, constant magnetic field at the same time as a stronger alternating field is allowed to decay (Jackson *et al.*, 1988). Alternatively the sample can be magnetised completely and then low coercivity fractions removed progressively by alternating field demagnetisation (Trindade *et al.*, 2001). The anisotropy of remanent magnetism within different coercivity intervals is called the partial anhysteretic remanent magnetisation. It has proved to be particularly useful for isolating fabrics produced by late secondary alterations (e.g. Trindade *et al.*, 2001).

5.6.3 *Other integrated (bulk) analytical methods*

Neutrons are diffracted by atoms and can therefore be used to determine the lattice parameters and orientation of crystals. Neutron beams cannot be focused; hence the method uses a broad beam that illuminates many crystals and gives bulk properties. The weak interactions between neutrons and most rocks mean that large samples must be used. Hence, the method is attractive for rocks with large grain sizes. Almost no sample preparation is needed: some workers shape their samples into 1 cm-diameter spheres (Xie *et al.*, 2003); others just use irregular blocks (Schafer, 2002).

Pulsed neutron sources enable the use of time-of-flight in addition to angular position information for analysis (Schafer, 2002, Xie *et al.*, 2003). Recent developments in software have enabled the method to analyse low-symmetry minerals such as plagioclase. The method is limited by the small number of neutron sources and detector arrays that can be used for these analyses. Sample analysis time is generally long and demand for these facilities intense, hence few samples can be analysed. Synchrotron X-rays can also be used and this may be the source of choice in the future (Wenk & Grigull, 2003).

X-ray diffractometry can also be used to determine the orientation of mineral grains in a limited class of materials. It is usually only practicable for minerals with an orthorhombic or higher symmetry. Mixtures of many minerals may also cause problems, as the diffraction lines cannot always be separated and identified. The low penetration power of X-rays in most rocks means that only the surface of a sample is examined. Despite these restrictions the low cost of the apparatus, when used on an existing X-ray diffractometer, can make the method attractive (Rodriguez-Navarro & Romanek, 2002).

5.7 Extraction of grain orientation data and parameters

The methods described above give a variety of different types of data: some techniques measure the bulk properties of the sample, whereas others give information on individual crystals in a rock. Individual crystal data are much richer but may not always be simple to interpret. Hence, individual data are commonly reduced to an orientation ellipsoid ($\equiv$ second rank tensor), which is very useful for treating large numbers of samples or comparing data with methods that measure bulk properties (see below and Mardia & Jupp, 2000). The orientation ellipsoid should be distinguished from the strain ellipsoid, which is a measure of the deformation required to produce the orientation ellipsoid from a population of randomly oriented lines. In a 2-D section the two are identical, but in 3-D they may differ: the directions of the axes are identical, but their relative lengths may be different (Gee *et al.*, 2004). If we have 3-D data, such as EBSD, then we can calculate the orientation ellipsoid directly from these data. However, many popular methods can only be applied to sections: here we must integrate data from several non-parallel sections to get the 3-D ellipsoid (see below).

5.7.1 Individual grain orientation display

Individual crystal data are commonly displayed graphically so that they can be interpreted in terms of geological processes. Three angular parameters are needed to express the three-dimensional orientation of each independent crystallographic (lattice) axis or shape axis of a crystal. This can only be expressed in a three-dimensional diagram, such as the orientation distribution function diagram (ODF) (Schmid *et al.*, 1981, Bunge, 1982). Because of the obvious difficulties of ODF diagrams geologists commonly use projections of polar diagrams of pole figures. The most popular is the equal-area net (Schmidt or Lambert net). SPO can be displayed as the directions of major and minor (longest and shortest) axes of each crystal with respect to specimen reference directions. LPO can be displayed in a similar way by plotting poles to lattice planes, crystallographic axes or zone axes with reference to some geologically relevant orientation. For uniaxial crystals one net gives most of the useful fabric information, but two nets are needed for the complete specification of the orientations of all crystals. Three nets may make some fabric symmetry more evident. In some studies the inverse pole diagram is used

(Randle & Caul, 1996). Here, the crystal axes are the reference frame and the orientation of the lineation plotted for each crystal.

Two-dimensional orientation data can be readily displayed as we require only two parameters. The simplest diagram is a conventional histogram of crystal numbers against angle. Data bins must be chosen with regard to the number of data points: more detail is possible if more data are available. A variation on this theme is the rose diagram, which is a polar histogram of the number of crystals against real angle. Such a diagram is symmetric, because crystal orientations do not have a distinct direction. Such a rose of orientations is a periodic function that can be analysed by the Fourier method. This can have as many terms as the number of data points, but for real data it is best to truncate it (Launeau & Robin, 1996). The size and shape of the data window analysed may be important if few grains are present. In this case edge effects must be considered; that is grains that lie across the boundary of the analysed area. The simplest solution is either to increase the area measured or to use a circular section. However, the latter assumes that the fabric is not strongly developed.

Orientation data can also be displayed without dividing into bins in the cumulative distribution function against angle diagram. This is a plot of the number of crystals with an orientation angle less than a certain value against the angle. It is easy to apply some statistical tests to this function as it increases monotonically from zero to n, the number of samples. It is also used by Gee *et al.* (2004) to combine sectional data into an orientation ellipsoid (see Section 5.7.3).

The variation in crystal orientation across a surface can be coloured according to a 'look-up' table. The table contains various bins with their corresponding colours. The colours are then applied to a map of the surface or volume. These diagrams are sometimes called AVA diagrams.

5.7.2 Mean orientation from individual data

The statistics of 3-D orientations are complex, but have been well developed and described (e.g. Mardia & Jupp, 2000). For AMS studies the accepted best way to calculate mean orientations is that of Hext (1963) and Jelinek (1978). Such statistical methods are commonly incorporated into the programs that reduce AMS data.

The statistics of 2-D data are simpler but present a few complexities, hence are briefly described here (see also review in Capaccioni *et al.*, 1997). We are usually interested in the mean and degree of dispersion of the orientation data. The mean grain orientation angle, $\bar{\Theta}$, cannot be calculated as a simple mean because the orientation variable is circular, that is $0° = 180°$. Note that this method is slightly different from that for direction data, which varies from 0 to 360°. The simplest method uses the sum of the direction vectors, with the angles doubled so that the data vary from 0 to 360°:

$$x_r = \sum \sin 2\Theta_i \text{ and } y_r = \sum \cos 2\Theta_i \text{ where } \Theta_i \text{ is the orientation}$$

of the *i*th outline.

$$\bar{\Theta} = \tan^{-1}(x_r/y_r)/2$$

Note that $\tan^{-1}$ must be corrected for the quadrant. The length of the sum of the direction vectors divided by the number of measurements can be used as a test of significance, or departure from random distribution, or can be a measure of the geological dispersion:

$$\bar{R} = \frac{1}{n}\sqrt{\left(\sum \sin \Theta_i\right)^2 + \sum (\cos \Theta_i)^2}$$

Statistical tables are available to test if the distribution is significant (Swan & Sandilands, 1995).

Another approach is to use the eigenvalue associated with the first eigenvector of the direction cosine tensor, T (Harvey & Laxton, 1980, Launeau & Cruden, 1998):

$$T = \frac{1}{N} \begin{vmatrix} \sum \sin(\Theta_i)^2 & \sum \sin(\Theta_i)\cos(\Theta_i) \\ \sum \sin(\Theta_i)\cos(\Theta_i) & \sum \cos(\Theta_i)^2 \end{vmatrix}$$

where N = number of crystals and Θ = orientation of crystals. The dispersion of the orientations is derived from the eigenvalues, e_1 and e_2. For no alignment $e_1 = e_2 = 0.5$; for perfect alignment $e_1 = 1.0$ and $e_2 = 0$. Several parameters are used: one is the ratio of the axes of the orientation ellipse $= \sqrt{e_1/e_2}$. For this measure massive rocks have a value of one and perfectly foliated rocks have a value of zero. Another is the alignment factor $= 2(e_1 - 0.5)$, (or 100 times this value). In this case massive rocks have a value of zero and perfectly foliated rocks have a value of one.

5.7.3 Integration of 2-D section data into a 3-D orientation ellipsoid

Some of the techniques described above give information on sections through a rock. Data from at least three non-coplanar sections are needed to give a three-dimensional orientation ellipsoid that summarises the data (Launeau, 2004). Ideally, one section should be parallel to the foliation, a second orthogonal to the foliation and parallel to the lineation and the third orthogonal to the lineation. However, if the foliation or lineation is weak then this is difficult.

The sectional ellipse method starts with the reduction of the data from each section to a 2-D orientation ellipse (see above). If simple orientation data are measured, such as crystal lattice or shape orientations, then the orientation of the major axis of the ellipse is clear but the absolute size is unknown: instead, we just know the ratio of the major and minor axes. Robin (2002) has developed a robust method that gives the best-fit ellipsoid from three or more sections and a quality-of-fit parameter (see Appendix). However, Gee *et al.* (2004) consider that this method does not give an accurate ellipsoid for strong fabrics, especially where the sections are not cut accurately along the significant planes of the ellipsoid.

The approach of Gee *et al.* (2004) is stochastic. They initially reduce the orientation data in each section to the cumulative distribution function (CDF; see above). The contribution of each measurement to the CDF can be weighted by the crystal length or area. They then take a population of randomly oriented vectors (representing the crystals), apply a strain tensor to their orientations (equivalent to deformation) and calculate a CDF for each section. The resulting CDFs are compared to the data CDFs for all the sections and the total misfit calculated. The parameters of the strain tensor are then varied to minimise the misfit between the CDFs. The orientation ellipsoid can be calculated from the strain tensor that gives the smallest misfit (see Appendix). The uncertainties in the orientation ellipsoid parameters were also calculated. The method was tested on two samples and appears to give an orientation ellipsoid that conforms closely to that calculated from individual crystal measurements by EBSD. It was much better than the sectional ellipse method for a sample with a strong fabric.

5.8 Typical applications

Applications were chosen because they illustrate what people are doing, or because they show interesting approaches. AMS has been used for a large number of studies in all branches of petrology. Recent collections include a special issue of *Tectonophysics* (**307**, 1–234), a book (Martin-Hernandez *et al.*, 2005) and a review article (Borradaile & Henry, 1997).

5.8.1 Flow in lavas

Although movements of the exterior of lava flows can be observed easily, it is generally necessary to determine the directions of movements within the flow, so that the exact mechanism of flow may be established. A number of models are possible: plug, Bingham or Newtonian flow, similar to flow in an enclosed conduit; glacier type flow; or rolling of the lava front beneath the flow. Some of these possibilities have been explored for various lava compositions.

5.8.1.1 Mafic and intermediate lavas

A short, crystal-rich dacite lava flow on Monte Porri, Salina, Italy was investigated by Ventura *et al.* (1996). A single detailed section through the central 2.2 m-thick massive part of the flow was examined with 19 samples. The authors measured tabular plagioclase crystals with axial ratios of ~4 and quantified their orientations and sense of imbrication from oriented thin sections. Crystals with lengths of 0.2–1.0 mm had orientations that varied consistently through the flow section: at the base and top the crystals were aligned parallel to the base and top with the strongest fabric. Towards the interior, the fabric became more inclined, up to 40°, and weaker. In the central part of the flow the fabric was not evident. Imbrication directions agreed statistically with the flow sense indicated by the crystal orientations, but 30%–40% of individual data were in the opposite direction. Crystals from 0.04 to 0.2 mm long were randomly oriented and appear to have grown in place after flow finished. The authors conclude that the textures accord with a plug flow model between two rigid walls. This is a bit surprising as the upper part of the flow is not confined. Maybe the scoriaceous breccia zone acts as a brake on movements within the flow or perhaps magma continued to move in the flow after the upper surface had solidified.

Many lava flows and plutonic rocks contain ellipsoid bodies of contrasting composition, which can also be used to determine flow directions and mechanisms. These enclaves result from mixing of magmas and it is important to determine if the enclaves have deformed internally during magma movement or if they have merely rotated as rigid bodies. A number of factors intervene, but the most important are probably the viscosity contrast with the host magma and the yield strength of the enclave and host magma (Ventura, 2001). The Monte Porri dacite lava flow was used as an example (Ventura, 2001) because crystal orientations had already been used to determine the flow regime (Ventura *et al.*, 1996). Enclave orientation and shape were measured in vertical sections near the base of the flow and used to establish the proportion of pure shear (compaction) to simple shear. Near the vent flow was hyperbolic, reflecting a mixture of spreading of the flow and gliding (Figure 5.6). The component of simple shear increased with distance along the flow indicating an increase in the gliding component.

Egmont volcano is a fairly typical stratovolcano covered with andesite flows. Most are thin, and have commonly almost drained away, but there are a number of much thicker flows. The bases of two flows were examined by

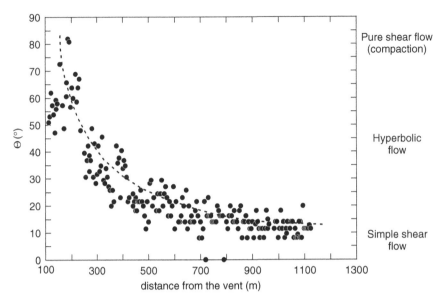

Figure 5.6　Nature of shear in the Monte Porri dacite lava derived from measurements of the orientation and shape of ellipsoid enclaves (Ventura, 2001). The angle Θ is the angle between the streamlines and the base of the flow. It is $90°$ for pure shear (compaction), $45°$ for hyperbolic shear and $0°$ for simple shear.

Higgins (1996a): the Kokowai Valley flow is about 10 m thick, and Humphries' Castle is over 100 m thick. The SPO of plagioclase was determined optically using thin sections (Figure 5.7). In both cases the preferred orientation direction showed a consistent pattern: near the base the crystals are oriented parallel to the slope of the flow, but above a height of 1–2 m the preferred orientation dips at about $30°$ in a downstream direction. This is opposite to the direction indicated in the Monte Porri dacite flow (Ventura *et al.*, 1996). The significance of the intermediate sample from the Kokowai Valley flow that dips upstream at about $20°$ is not clear. The quality of plagioclase orientation does not appear to correlate strongly with sample position. These orientations suggest that the flow advanced with a rolling motion: crystals were aligned during flow further upstream and then rolled over the stagnant zone close to the base of the flow. However, more data is needed to confirm this.

Traditionally, only the surface topology of lava was used to classify the flow as a'a, pahoehoe or toothpaste. However, the differences between these types are more fundamental and continue into the interior of the flows: differences in CSDs have been noted in Section 3.4.1.2. The transition from pahoehoe to toothpaste to a'a is related to both the melt viscosity and the strain rate.

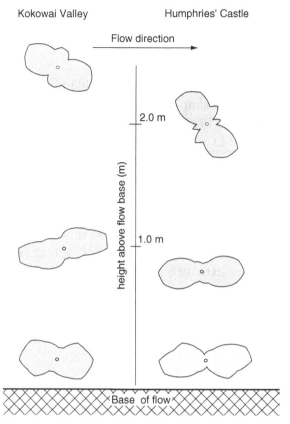

Figure 5.7 Orientations of plagioclase crystals in basaltic andesites from Egmont Volcano (Mt Taranaki), New Zealand (Higgins, 1996a). Two different flows were sampled, each at three heights. SPOs are displayed as rose diagrams.

Canon-Tapia *et al.* (1997) investigated these differences in a series of basalt flows from Hawaii using AMS. They surmised that the magnetic characteristics of the lavas can be explained by the distribution of the magnetic particles in the lava and post-emplacement crystallisation. However, the orientation and degree of anisotropy of magnetic susceptibility is defined by shear of the magma during consolidation. Although some authors have suggested that k_2 is parallel to the flow direction, Canon-Tapia *et al.* (1997) consider that here it is k_1. In all the flows the shear rate was found to be most intense at the base of the flow, as is to be expected. However a'a flows were in general more sheared than pahoehoe flows. The directions of AMS within the flows seem to be distributed more regularly in a'a as compared to the pahoehoe flows. This suggests that a'a continues to flow until it is stopped by high viscosity. In contrast, pahoehoe flows appear to cease flowing when the magma is still fluid. This enables late flow and/or endogenous growth after the flow has stopped and dispersed AMS directions.

5.8.1.2 Felsic lavas

Some microlites are prismatic and hence provide almost ideal markers for determining the amount of simple shear and pure shear (compaction) in lava flows and domes (Manga, 1998). Some glassy felsic volcanic rocks are transparent enough that the 3-D orientation of microlites can be measured directly. Obsidian Dome, USA, is a rhyolite flow about 3 km is diameter: orientations of pyroxene microlites show that the flow developed by radial flow followed by flattening: pure shear increases from 0.3 at the vent to 1.1 at the flow front (Castro *et al.*, 2003).

AMS measurements were also made on the same rocks (Canon-Tapia & Castro, 2004). Although the bulk susceptibility of the rocks is low the degree of magnetic anisotropy is extremely high. Directions of the maximum axis accorded with the lineation direction of the microlites and showed the local direction of flow. The authors caution that their results may not be applicable elsewhere if the susceptibility is not caused by the same minerals with similar prismatic forms.

5.8.2 Flow in dykes

The textural fabric of dykes is of immense importance as it can be used to determine the direction of flow of the magma (Blanchard *et al.*, 1979). Hence, it is possible to locate the position of the original magma chambers, which can in turn tell us about regional geodynamics. Many studies have used AMS to determine flow directions (e.g. Ernst & Baragar, 1992), but there are significant problems in the interpretation: it is not always clear how the magnetic vectors correspond to the magmatic deformation field. In some cases k_2 appears to be parallel to flow direction, whereas in others it is k_1 (Geoffroy *et al.*, 2002). There are a number of ways in which this uncertainty can be resolved.

Optical methods can be used to determine the flow direction of magma in phyric dykes. Oriented thin sections are cut from lateral positions on either side of the dyke and the imbrication of magmatic foliation can be used to determine the flow direction (Figure 5.8). In one of the earliest studies a section was cut parallel to the dyke walls and the orientation of tabular plagioclase crystals determined using a universal stage (Shelley, 1985). Such LPO measurements are very tedious hence more recent studies tend to examine only the SPO of phenocrysts. This is readily done directly from a thin section using the

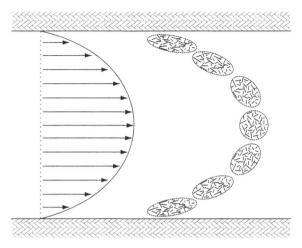

Figure 5.8   Flow of magma with crystals (or enclaves) in a dyke. Initially randomly oriented crystals develop a strong fabric near the dyke walls that angles into the flow centre towards the direction of flow. This imbrication is mirrored on the other side of the dyke. Fabric is not developed in the centre of the dyke.

intercept method (Geoffroy *et al.*, 2002). In one study of a composite dyke the directions of emplacement of the three separate phases could be identified (Wada, 1992). In some studies the flow direction was determined by optical methods and used to establish the axis of the corresponding AMS ellipsoid. The AMS could then be applied to a much larger group of samples (e.g. Poland *et al.*, 2004). Another possibility is to look at the imbrication of AMS ellipsoids from samples taken on either side of the dyke, a process analogous to the optical determination (Geoffroy *et al.*, 2002).

Mafic dyke swarms can cover huge areas and individual dykes can sometimes be traced for hundreds of kilometres. Ernst and Baragar (1992) examined the McKenzie dyke swarm, which radiates from a point in northern Canada and extends southward for 2000 km. Such dyke swarms are commonly associated with mantle plumes and it is of interest to determine the original area of magma production. Ernst and Baragar (1992) showed that close to the focus of the swarm the flow direction has an important vertical component, but this becomes horizontal away from the focus. This shows that mafic magmas can be propagated over vast distances in the crust.

Magma flow directions in a dacite dyke on a more local scale were examined by Poland *et al.* (2004). Two dykes 3 km long were investigated using AMS with verification of the flow directions from measurements of crystal orientations. The flow directions in the dykes were simple and linear in their deeper parts, but broke up into en echelon segments close to the surface (Figure 5.9).

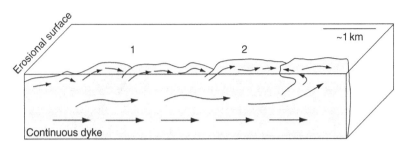

Figure 5.9   Emplacement mechanism of a dacite dyke derived from thin
section analysis and AMS measurements (Poland *et al.*, 2004).

The dominant flow direction in the deeper parts of the dyke was horizontal.
Flow in en echelon segments is generally assumed to be parallel to the axis of
segment rotation. However, this was not found to be generally the case here:
again dyke flow was largely horizontal, either parallel with the overall direction of flow, or into the tips of the segments in the opposite direction.

Although many dykes form by simple extension normal to the walls, dykes
are a zone of weakness and hence can easily accommodate lateral displacement
of the walls, as in a fault. In narrow dykes such displacement can commonly be
seen in the field, but it is not so evident for larger dykes. However, displacement can be deduced from AMS measurements of fabric obliquity near the
walls of the dykes. In simple dykes fabric obliquities are opposed on either side
of the dyke (Figure 5.8). If there has been displacement of the walls during
solidification then the obliquity on one side may be inversed, making it parallel
to that on the opposite side. Scott and Benn (2001) examined dykes radial to
the Sudbury impact structure. Their AMS measurement revealed a steep
foliation that was counterclockwise oblique on both sides of the dyke. They
proposed that the dyke acted as a magma-filled transfer fault during collapse
of the inner ring of the central peak, all formed just after the impact of the
meteorite.

The magnetic fabrics of altered and metamorphosed dykes may be complex
and difficult to understand. Borradaile and Gauthier (2003) examined dykes
from the Troodos ophiolite complex, Cyprus, which had been altered by
hydrothermal fluids. Original magnetite and titanomagnetite were partly
replaced by the more oxidised mineral maghemite. They found that maghemite
was a significant component of the bulk susceptibility, but that AARM
indicated that this secondary magnetic fabric had a low anisotropy. Hence,
AMS measurements corresponded to magmatic flow directions. Clearly, this is
not always the case and AARM determination are useful to decode anomalous
magnetic fabric directions.

5.8.3 Magmatic deformation in plutonic rocks

5.8.3.1 Mafic rocks

Mineral layering and foliation (lamination) is widely observed in mafic plutonic rocks but most studies have been qualitative, rather than quantitative (Wager & Brown, 1968, Naslund & McBirney, 1996). The upper part of the Sept Iles intrusive suite contains anorthosite some of which is very well foliated (Higgins, 1991). The shape and SPO of plagioclase in a foliated sample was determined using orthogonal thin sections. Higgins (1991) showed that a foliation of such quality could only be produced by simple shear whilst there was still silicate liquid between the grains. The lack of an observable lineation may reflect the shape of the tablets: they are almost equant in the plane of the largest surface (1:5:7).

The Kiglapait mafic intrusion, Labrador, Canada, has been well studied as it is undeformed, relatively small and well exposed (Morse, 1969). The lower zone is troctolite and the upper zone is gabbro. Plagioclase and olivine SPOs have been determined both for a section through the intrusion and a detailed study of a single layer (Higgins, 2002b). Single samples were orientated vertically directed towards the centre of the intrusion. SPOs were measured optically and reduced using the inertia tensor method.

Plagioclase SPOs are significant throughout the section, but are strongest in the lower quarter of the intrusion (Figure 5.10). Simple compaction is not able

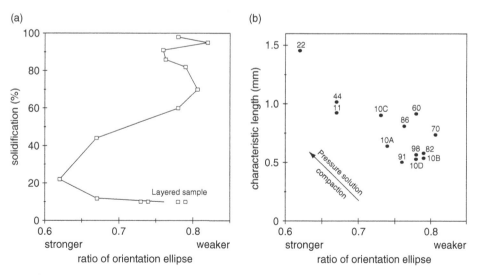

Figure 5.10   Plagioclase SPO in the Kiglapait intrusion, Canada (Higgins, 2002b). The ratio of the orientation ellipse indicates the quality of the foliation: 1 = massive; 0 = perfect foliation.

to produce significant foliation (Higgins, 1991): hence the strength of the SPO must reflect the vigour and stability of the flow within the chamber. Well-layered and foliated rocks near the base of the intrusion may represent a period of stagnation interspersed by the flow of crystal laden turbidity currents. The more strongly foliated samples higher up indicate a period of stable convective flow, which diminished towards the upper part of the intrusion. Olivine also has a significant SPO in all samples, which is not surprising as olivine is also a liquidus phase (Morse, 1979).

As was mentioned before compaction alone cannot produce a significant SPO, unless the crystals change shape by solution and growth (Hunter, 1996, Meurer & Boudreau, 1998). Such a process is related to both pressure solution and coarsening. Hence, we would expect to see a correlation between the strength of the foliation and the characteristic length. Such is observed for two samples from Kiglapait, suggesting that this process is either only locally important or only locally preserved.

The Tellnes deposit, Norway, is a large lens of ilmenite-rich norite within a long jotunite dyke. Measurements of the petrofabric by AMS and image analysis have been used to determine the mode of formation of the deposit (Figure 5.11; Diot *et al.*, 2003). Partial anhysteretic remanent magnetism was used to show that the AMS was produced by coarse magnetite grains. The significance of magnetic fabrics is not always clear, hence the fabric of some

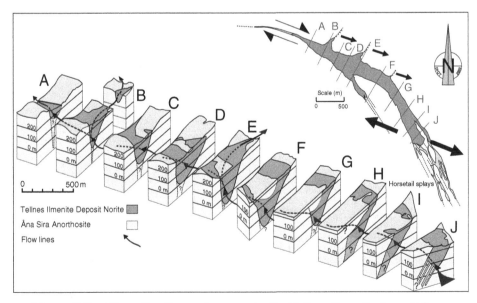

Figure 5.11   Flow directions of magma in the Tellnes ilmenite deposit from AMS measurements (Diot *et al.*, 2003).

samples were also determined using optical methods. Three orthogonal thin sections were cut and the fabric produced by both ilmenite and magnetite was quantified using the intercept method. In general the fabric orientation determined from AMS and image analysis agreed within 30°, despite the fact that AMS looks at magnetite whereas the optical methods mostly reflect ilmenite distributions. The fabric distributions suggest that the deposit was formed by multiple injections of crystal-rich mushes and that the fabric was developed by magmatic deformation during emplacement.

The lower part of the Bushveld complex, South Africa, is comprised of rocks rich in orthopyroxene. Boorman *et al.* (2004) studied the alignment of these prismatic crystals as part of a comprehensive textural study. They found that there was a good correlation between the aspect ratio of the crystals and their degree of alignment. However, there was no correlation between the degrees of alignment normal and parallel to the foliation, indicating that there was no lineation. If the foliation was produced by simple shear (flow) then such prismatic crystals should have a significant lineation, unlike the tabular plagioclase crystals discussed above. They conclude that the mineral alignment was produced by pure shear (compaction) accompanied by coarsening. This process is more efficient in rocks dominated by orthopyroxene as the presence of other phases will pin the grain boundaries. It is also similar to the pressure solution compaction proposed by Meurer and Boudreau (1998) and seen sporadically in the Kiglapait intrusion (Higgins, 2002b).

The Oman ophiolite has been the subject of numerous studies as it is well exposed and relatively unaltered. Yaouancq and MacLeod (2000) studied the fabric of the gabbros using field measurements, AMS and universal stage methods. They found that the AMS only followed the megascopic foliation or LPO in about 2/3 of the foliated gabbro samples. In these rocks the AMS is controlled by secondary magnetite produced by alteration of olivine along fractures. In many samples these fracture planes follow the overall fabric. However, there is no correspondence between the AMS and plagioclase LPO ellipsoids. Gabbros from the top of the section show no link between AMS and overall rock fabric: here AMS is produced by primary interstitial magnetite grains whose distribution is not controlled by the rock fabric. Hence, application of AMS on these types of rock is extremely problematical and it is best to quantify rock fabric using measurements on mineral orientations, such as plagioclase.

Compaction of cumulate rocks is thought to be an important process during solidification, but has rarely been quantified. The recent discovery of plagioclase crystal chains in the Holyoke basalt flow and other units has provided a structure that can be quantified and used to determine the degree of compaction (Philpotts *et al.*, 1998, Philpotts *et al.*, 1999). Grey *et al.* (2003) examined the texture of a sample of gabbro from the Palisades sill, New Jersey, using four different techniques. They first traced the plagioclase crystal chains from thin section photographs. From these line networks they determined the anisotropy using the intercept method (see Section 5.4.3.2). They also determined the anisotropy of the chains from the orientation and size of the links between adjacent crystals and the geometry of the interstitial regions between the chains. All these methods produced a similar measure of anisotropy which was interpreted as a result of compaction of 8%. In addition the anisotropy plunges down a dip, suggesting that the sill was emplaced in its present orientation and that there has been minor deformation of the mush in this direction, consistent with the compaction.

The anisotropy was also determined independently of the plagioclase chains directly from grey-scale thin section photographs. Images of the silicate and opaque minerals were analysed using the directional variogram method. The silicate minerals yielded a similar direction and shape of anisotropy to the data from the plagioclase chains. However, the opaque minerals gave a very different direction of anisotropy, orthogonal to that of the silicates. This was similar to that provided by AMS analysis of the same material. This suggested that the opaque minerals crystallised late, perhaps along cracks that opened up during compaction.

5.8.3.2 Granitoid intrusions

Granitoids are a major igneous component of the middle and upper crust and were commonly emplaced during deformation events. Numerous studies of the fabric of granitoid intrusions show that it is remarkably homogeneous, or varies smoothly through an intrusion and hence it records the deformation that the magma underwent during solidification (Bouchez, 1997, Bouchez, 2000). Some authors consider that the deformation was produced by flow in the magma chamber during emplacement, whereas others have shown that the fabrics record regional strain during the last stages of solidification.

The fabric of granitoids has been extensively examined using AMS, sometimes supplemented by mineral orientation studies (Bouchez, 1997, Bouchez, 2000). If magnetite is present, then it will completely dominate the overall susceptibility of the rock. Gregoire *et al.* (1998) examined a magnetite bearing quartz-syenite and found a good correlation between the magnetite shape fabric and the AMS ellipsoid. This contrasts with more mafic rocks, in which the relationship has not always been so simple (e.g. Yaouancq & MacLeod, 2000). In the absence of magnetite the AMS is dominated by the shape fabric of paramagnetic minerals, such as biotite and amphibole. If these minerals were present when the rock was deformed, then the AMS and mineral fabrics will be identical.

The Mono Creek pluton is a good example of a magnetite dominated system (Bouchez, 2000). It was emplaced into earlier components of the Sierra Nevada Batholith, California, during a period of regional dextral transgression. The AMS does not follow the border of the intrusion, as would be expected if fabric were due to convection or other internal processes (Saint-Blanquat & Tikoff, 1997). Instead, the AMS defines a broad sigmoidal shape that accords with emplacement during tectonism. Towards the end of solidification the deformation was localised and became more intense, producing a stronger AMS fabric. Deformation continued after solidification with the production of a shear zone.

In many granitoids rock fabric, as expressed by AMS, clearly indicates a sense of deformation, but this deformation may be a late event, unrelated to the main phase of the emplacement of the intrusion. Cruden and Launeau (1994) have investigated the Lebel syenite intrusion, Ontario. AMS fabrics, defined by magnetite, indicate that the intrusion has a sheet-like form. They propose a model in which the magma rose up an E–W fault and was then injected horizontally as a tongue of magma into the host rocks. Launeau and Cruden (1998) extended this study by the analysis of mineral fabrics using the intercept method. They found that the mineral fabrics were generally parallel to the AMS ellipsoids: however, in some samples bimodal mineral fabrics were attributed to overprinting or complex movements of grains with different shapes. The experiments of Iezzi and Ventura (2002) confirm that minerals with different shapes can behave differently during magmatic deformation. Crystal size distributions of pyroxene and magnetite were straight and this was interpreted to mean that these minerals crystallised over a protracted interval. Hence, the AMS was mostly produced during the main phase of crystallisation of the body.

The fabric of some granitoids has been investigated using optical methods at the same time as other aspects of the texture were quantified. The Halle Volcanic Complex, Germany, is a porphyritic rhyolite intrusion with abundant quartz, orthoclase and quartz phenocrysts (Mock *et al.*, 2003). Optical measurement of sections, some of them orthogonal, consistently showed significant preferred orientations at 45° and 135°. This was interpreted as conjugate zones, developed during crystal settling. However, the consistent direction for both vertical and horizontal sections, and the similar, weak orientations, suggests that it may just be an artefact of the measurement process.

5.8.3.3 Engineering studies

The mechanical properties of rocks are of fundamental importance to engineers and there have been a number of attempts to link this to rock texture. Early studies used a single parameter to summarise the effects of texture on rock properties (Howarth & Rowlands, 1986). The 'texture coefficient' is derived from the orientation and shape of crystals and porosity of the rock, but does not seem to be a good predictor of rock properties (Ersoy & Waller, 1995). Akesson *et al.* (2003) examined the effects of rock fabric and texture on the mechanical properties of rocks. They used the intercept method to quantify rock foliation and found that there was a good correlation between the degree of foliation and the aspect ratio of the fragments. In a previous study of isotropic rock the same group had found a simple relationship between the area-normalised grain perimeters and rock fragility as expressed as the Los Angeles (LA) index (Akesson *et al.*, 2001). The foliation index that they developed can be used to correct the normalised perimeter to that of isotropic rocks, and hence extend the use of this relationship.

5.8.4 Solid-state deformation

It has been known for some time that the upper mantle is seismically anisotropic, which is caused by LPO of olivine and orthopyroxene. Hence, the tectonics of the mantle can be determined by remote measurements of its fabric. Ben Ismail and Mainprice (1998) have investigated the nature of the link between fabric and seismic anisotropy using laboratory measurements of fabric in mantle samples. They examined 110 samples from a wide variety of settings and with a range of textures. They classified the samples into four groups on the basis of the orientation of their olivine lattices with respect to lineation and foliation directions. In all cases the [100] direction was parallel to the lineation and this had the greatest influence on the P-wave anisotropy. The [010] and [001] directions were much more variable and provided the key to the fabric classification. It was found that only the [100] and [001] directions influenced the S-wave anisotropy.

Garnet is a common mineral in metamorphic rocks, but its isometric symmetry hides textural detail in optical studies. However, garnet is not structurally isotropic and hence measurement of lattice orientations can potentially reveal petrogenetic processes. Prior *et al.* (2002) used EBSD to examine two garnet grains from two different rocks: the Glenelg was a deformed grain and the Alpine was a porphyroblast (Figure 5.12). Each grain comprised many sub-grains (defined as an angular difference $>2°$) and they compiled the

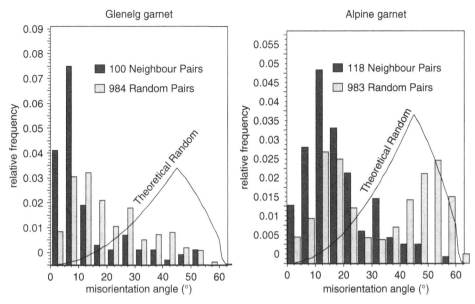

Figure 5.12   Orientation of garnet sub-grains by EBSD (Prior *et al.*, 2002). The sub-grains in the Glenelg garnet were produced by solid-state deformation, whereas those in the Alpine garnet are the result of coalescence of independently nucleated grains, followed by rotation.

angular difference (misorientation) between adjoining and randomly selected sub-grains. These distributions were compared with that expected if the sub-grains had random orientations. In both cases adjacent grains had lower misorientations that randomly selected grains. In the deformed garnet the sub-grains were internally distorted and the misorientations lay along small circles around rational crystallographic axes. Hence, this the subgrains likely formed by deformation of an original large grain. In the porphyroblast sub-grains are undistorted and misorientations have random orientations. This grain formed by coalescence of neighbouring grains. However, misorientations are lower than expected for such a process and it must have been accompanied by sub-grain rotation (Spiess *et al.*, 2001). This would serve to minimise the total surface energy of the grain, as do grain-boundary migration (coarsening) and adjustments to the dihedral angles of the crystals (see Sections 3.2.4 and 4.2.3).

5.8.5 Fabric development in pyroclastic rocks

The products of large explosive eruptions can cover vast areas with readily identifiable pyroclastic deposits. However, the vent area of these deposits is not always easy to identify, as it may have been removed or filled in by later eruptions. Image analysis has also been used to establish the orientation of the fabric in tuffs (Capaccioni & Sarocchi, 1996). Blocks were imaged and the clasts electronically separated. The fabric parameters were derived from the individual grain outlines. However, sedimentological structures may not be strong enough to be identified and cannot always be applied, for instance in drill core. In some areas AMS has been used to identify the flow directions of such tuffs.

AMS has an orientation, but not a direction. Hence, if the major axis of the AMS ellipsoid (k_1) follows the flow lines of the tuff we can get two possible directions. Luckily, in many tuffs the AMS ellipsoid dips up-flow at angles of 10–30° (Knight *et al.*, 1986, Palmer & MacDonald, 1999). Studies of recent undeformed tuffs have shown that the method can be used to identify vent regions. The degree of anisotropy does not correlate with density or degree of welding, indicating that the AMS was produced as a consequence of flow and not afterwards (Palmer & MacDonald, 2002).

Notes

1. The term fabric is used here for the orientation of grains in a rock. The equivalent term in materials science is texture or microstructure.

6

Grain spatial distributions and relations

6.1 Introduction

Grains in many rocks are not distributed randomly in space, but organised into clusters, layers and chains (Figure 6.1). Such spatial patterns can be defined in a number of ways: by the presence of grains as well as by their size, shape, orientation and mineral associations. Such non-random distributions of grains, termed patterns or packings, have been observed in metamorphic, igneous and sedimentary rocks (Rogers *et al.*, 1994, Jerram *et al.*, 2003). Despite the fact that quantitative investigation of such structures in a rock was started in 1966 (Kretz, 1966a), there have been relatively few quantitative studies since that time. Many terms have been used to describe nonuniform grain distributions: in igneous rocks grain clusters are referred to as clumps, clots or glomerocrysts, whereas in metamorphic petrology other textural terms are used. Another important textural aspect of a rock is the spatial association of minerals, or lack thereof. This may reflect nucleation, growth or mineral breakdown processes, but has been much less studied than the actual crystal positions.

Patterns can be imposed on a rock during its formation by externally varying conditions or can develop spontaneously. For instance, during deposition of sediments variations in the flow regime or sediment source can produce layers; these are clearly imposed externally. Some layering in igneous rocks may result from similar sedimentary processes (Naslund & McBirney, 1996). Layering in metamorphic rocks can reflect layering of a sedimentary protolith, and deformation of dykes and enclaves. Layers or other structures are referred to as templates (Ortoleva, 1994). In some homogeneous materials patterns can arise spontaneously, or existing templates can be amplified and modified. These effects are called 'geochemical self-organisation' (Ortoleva, 1994). A classic example of this effect is Liesegang layering. Hence, grain clusters,

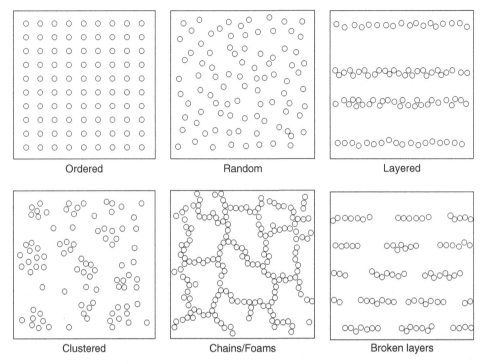

Figure 6.1 Schematic grain spatial distributions in two dimensions. Such patterns may be defined by the presence of grains of a phase, their orientation and mineral associations or by differences in grain size.

chains and layers appear to be common phenomena, with varying origins, and should yield petrogenetic information.

6.2 Brief review of theory

6.2.1 Nucleation of crystals

Nucleation occurs when it is energetically favourable to create a new phase, or more crystals of an existing phase (see Section 3.2). If nucleation occurs spontaneously within a phase it is termed homogeneous nucleation. The position of the nuclei will be random, provided that the temperature and composition of the liquid are constant. If, however, a region of the liquid is depleted in the crystal components, for example by crystallisation of an adjacent crystal, then nucleation may be suppressed there. This may produce megascopic patterns by geochemical self-organisation (Ortoleva, 1994).

Nucleation can also occur on existing phases, termed heterogeneous nucleation. The position of the nuclei of the new crystals will obviously depend on the location of the host materials. It is unlikely to be random, unless the host grains are very small and initially randomly distributed. It is more likely that the nuclei will be clustered. If the crystals do not move significantly during solidification, this may be visible in the final rock. It is debatable if any magma is ever completely free of crystals, hence heterogeneous nucleation may be the normal situation.

Minerals can also form by the breakdown of other solid phases. If two minerals are formed by such reactions in the solid state then they will be spatially associated.

6.2.2 Geochemical self-organisation

Many periodic geological structures are thought to originate by variations in externally controlled parameters. For example, sedimentary layering may form by variations in the mineralogy and/or size of the particles being deposited. The variation in externally controlled parameters is considered to be a 'template' which forces the structure on the rocks. However, patterns can also arise spontaneously without the intervention of a template: this vast subject is termed geochemical self-organisation (Ortoleva, 1994, Jamtveit & Meakin, 1999). A number of patterned structures in geology may be produced by this effect (Ortoleva, 1994):

- layers and spots in igneous rocks
- metamorphic differentiation and layering
- oscillatory zoning in crystals
- Liesegang banding and layering
- diagenetic patterns
- geodes, concretions, agates and orbicules

Geochemical self-organisation may seem to violate the second law of thermodynamics: that all systems evolve spontaneously towards a state of increased disorder. However, although geochemical self-organisation is clearly a local increase in order, it must be compensated by a greater decrease in order of the whole system.

Self-organised patterns originate by the dissipation of free energy and are hence called dissipative structures. The initial unordered system is subject to random variations of thermal origin. This 'noise' can be decomposed into periodic fluctuations of all possible symmetries, in the same way that any wave can be Fourier transformed into an infinite number of harmonic waves. When the system is close to equilibrium none of these patterns dominates and the system appears completely unordered. However, when the system evolves away from its equilibrium state certain patterns may be amplified by positive feedback loops and an ordering developed. If the structure is observable and stable it is termed a dissipative structure. Such patterns may be comprised of layers, spots, spirals, etc. The patterns form by reaction–transport: this involves solution of some grains, transport of material and growth of other grains. Some patterns are only maintained by a continuous supply of free energy; however, the occurrence of dissipative structures in old rocks indicates that they can be frozen in place and observed long after the activity of the system has ceased. It should be emphasised that dissipative structures can form only in systems that are maintained away from equilibrium.

A classic example of patterning is Liesegang banding. In the simplest situation this consists of bands richer in a phase in a two-dimensional system. Three-dimensional Liesegang layers also occur, but have been little studied (Boudreau, 1995). Two different origins have been proposed (Chacron & L'Heureux, 1999). In the pre-nucleation model, nucleation and growth of a crystal depletes the surrounding area in that component, inhibiting nucleation of further crystals. As the reaction front moves away, higher concentrations are encountered and nucleation is enabled again, starting the production of a new layer. However, Liesegang bands can also form after nucleation from a homogeneous material by competitive particle growth and coarsening (see Section 3.2.4). If a small region has a local increase in crystal size, then the concentration of the component in the liquid will decrease. As material diffuses in from adjacent regions the concentration there will decrease and hence the crystals will dissolve. The process can produce a series of bands with different crystal sizes and abundances. Both processes have been combined in a one-dimensional model by Chacron and L'Heureux (1999). However, observations of the actual formation of Liesegang bands and layers suggest the post-nucleation model is more important (John Ross, quoted in Boudreau, 1995). The post-nucleation process has been proposed for the origin of the famous 'inch-scale' layering of pyroxenes in anorthosite in the Stillwater intrusion (Boudreau, 1995). Philpotts and Dickson (2002) have observed millimetre-scale layering in the roof zone of a thick basalt flow. It consists of sheets of ophitic plagioclase–pyroxene clusters separated by sheets of residual liquid. They suggest an origin by cycles of nucleation and that such crystal sheets are separated by gravity along planes of weakness. It is also possible that the whole structure originates from self-organisation. I have observed that in some igneous rocks well-developed layering passes laterally into broken layers and finally clusters. Such behaviour is also predicted by self-organisation processes (Krug *et al.*, 1996).

6.2.3 Grain aggregation and agglutination

Ideally, a grain will sink or rise in a Newtonian fluid at a speed that is related to the density contrast and size by Stokes' law. However, in most geological situations the concentration of grains is sufficiently high that grains interact and will gather into clusters, if there is sufficient time (see the review in Schwindinger, 1999). Such clusters will move as if they were made of a single, larger grain, and will hence descend faster than each of the original grains. The clusters can form by vertical and horizontal grain convergence: a large grain will descend faster than smaller grains and sweep such grains from in front of it. Grains may move laterally to reduce differences in liquid velocity on either side of the grain – this is similar to the 'Bagnold effect' that produces a concentration of particles in the centre of a conduit. Grain aggregates may be preserved in volcanic rocks, but it seems unlikely that such structures can be observed in cumulate plutonic rocks.

In some aggregates euhedral crystals touch along crystal faces (Vance, 1969). One of the best-known examples is that of olivine crystals from the 1959 Kilauea Iki eruption (Figure 6.2; Schwindinger & Anderson, 1989). Most crystal pairs are united by faces with the same indices, and larger faces are more commonly in contact than smaller ones. The whole structure resembles a twinned crystal except that it is not a growth form. The term synneusis ('swimming together') has been used to refer both to nonoriented aggregates of crystals, and such aligned crystal groups. Schwindinger (1999) considered that such aligned groups cannot agglutinate during settling, but must be produced during shear flow or turbulence. The driving force of this process is the reduction of the total surface (interfacial) energy of the mineral grains: contacts

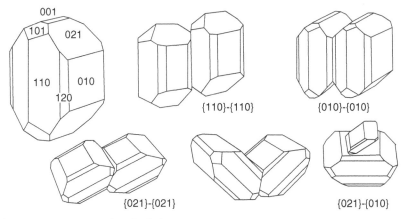

Figure 6.2   Synneusis of olivine grains from the 1959 Kilauea Iki eruption (from Schwindinger 1999).

between similar grains will have a lower surface energy than crystal–magma contacts. This process has also been proposed for the origin of some garnet porphyroblasts in metamorphic rocks (Prior *et al.*, 2002).

6.2.4 Magma mixing

If several magmas or crystal mushes are introduced into a conduit or chamber they will have a tendency to mix. The vigour and duration of the mixing will determine if the resulting material appears to be homogeneous on a microscopic scale or if the two magmas can still be recognised in blocks and outcrops: the two cases are commonly referred to as magma mixing and magma mingling. If the original magmas differ in their crystal contents then the resulting magma may have a non-random distribution of crystals. Of course, measurements of crystal positions must be done at a higher resolution than that of the expected structure. In some rocks the results of mixing are clearly revealed by compositional or isotopic studies. However, if mineral compositions are constant for all grains, then an origin of crystal clusters by incomplete mixing may not by easy to determine.

6.2.5 Grain packing

Grains in some sedimentary rocks, and crystals in some cumulate igneous rocks, are packings of hard particles (Rogers *et al.*, 1994). Theoretical packing of different sized and shaped particles is very complex; hence this subject has been explored using numerical models of simple shapes, commonly spheres. These studies have been concerned mainly with the overall packing density, that is its efficiency, rather than the actual positions of the particles. Jerram *et al.* (1996) used one such study to establish a reference texture for nearest neighbour distances.

6.3 Methodology

The position of a grain has several measurable parameters.

- The simplest is its centre, which can be approximated by the centre of mass and specified by a single triplet of x, y and z values.
- The position of the surface of the grain must also be considered: for instance, two large grains may be distant, but touching. Many more data are needed to specify the grain surface – many x, y, z triplets.
- The nature of a grain's neighbours: certain mineral pairs may be touching or spatially associated more frequently than would be expected for a random texture.

Most analytical methods assume that the sample is homogeneous. It is particularly important to avoid materials like vesicular lavas, which do not meet this criterion.

6.3.1 Three-dimensional analytical methods

The centre and surface of a grain can be determined precisely using three-dimensional methods, if the surface of the grain can be separated from adjoining grains by physical or analytical means. In the earliest study of crystal positions a sample was dismantled using a chisel and the crystal positions measured with a ruler (Kretz, 1969). However, such a laborious study has not been repeated. The extent of individual grains can be determined by serial sectioning, if their concentration is low enough that few crystals are touching, or if grain boundaries are clear (see Section 2.2.1). In transparent materials, such as volcanic glasses, the extent of crystals can be measured directly with a regular or confocal microscope (see Section 2.2.2). X-ray tomography can be used to determine the extent of grains if adjoining grains can be separated (see Section 2.2.3).

These direct methods are the only ones that can be used for grains with complex shapes. They are also the only ones that can be used to identify 'chains' of crystals (e.g. Philpotts *et al.*, 1999). However, it is not always possible to separate grains using current 3-D methods. Some two-dimensional methods are much better at separating adjoining crystals and hence may have to be used (see Section 2.7).

6.3.2 Section and surface analytical methods

In a section the centre of the grain intersection is not the true centre of the grain, and is therefore referred to as the centroid. For simple shapes the centroid is the projection of the centre of the grain onto the section. This is not the case for grains with complex shapes and they cannot be examined successfully in two dimensions.

Grain outlines can be acquired easily from rock surfaces, slabs and sections, using a variety of techniques described in Section 2.5. Such images can be processed manually or automatically to produce classified mineral images (see Section 2.6). Some methods require that the grain outlines are obtained from the classified images. The centroid of the grain intersection can be determined as the centre of mass.

If two grains touch in a section, then clearly they touch in three dimensions. However, it is not possible to determine from a section if adjacent, but separate, grains actually touch out of the section plane.

6.3.3 *Extraction of spatial distribution parameters*

6.3.3.1 *Nearest-neighbour distance ('big' R)*

One of the simplest ways of describing spatial distribution patterns uses the mean nearest-neighbour distance calculated from the centres of the grains (Jerram *et al.*, 1996, Jerram *et al.*, 2003). This is compared to the mean nearest-neighbour distance from a random distribution of points with the same population density. This is a method that can be applied to any 2-D textures where a grain or crystal can be defined by a grain centre co-ordinate. In principle, the method could be extended into 3-D, where the 3-D nearest neighbour is compared to a random 3-D distribution of points, but as yet very few 3-D data-sets of rock textures exist and this has not been tested. However, it is possible to compare the 2-D spatial distribution pattern to existing known reference textures where their 3-D spatial pattern is known (Jerram *et al.*, 1996, Jerram *et al.*, 2003). So far this has been limited to isotropic, homogeneous materials and it is not clear if strongly foliated rocks, or sections with a significant porosity, can be analysed in the same way.

The positions of grain centres are generally determined when the other textural parameters are determined. A simple procedure goes through the list of N crystal centroid positions and finds the distance to the nearest neighbour, r (see Appendix). The mean nearest-neighbour distance is

$$r_A = (1/N) \sum r$$

The mean nearest neighbour for a random distribution of points with the same population density (number of crystals per unit area), N_A

$$r_E = 1/(2\sqrt{N_A})$$

The ratio of these parameters, 'big' R describes the distribution of the grains.

$$R = r_A/r_E$$

In principle, for a distribution of random points $R = 1$. For clustered points $R < 1$ and for ordered points $R > 1$. However, crystals are not points: each occupies a certain volume in space and area in a section. Hence, crystals cannot be coincident like points. To enable the comparison of 'big' R values Jerram *et al.* (1996) calculated the random sphere distribution line (RSDL, Figure 6.3). This represents the 2-D 'big' R values calculated from 2-D sections taken through 3-D random distributions of spheres with different porosities up to the minimum porosity: that for random close packed spheres with a porosity of 37%. The RSDL is calculated using a combination of computer

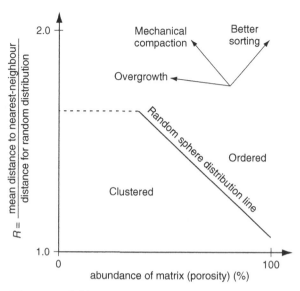

Figure 6.3 Nearest neighbour distance determined from sections through materials (Jerram *et al.*, 1996). The porosity is equal to 100 − abundance in % of the phase analysed. The random sphere distribution line is indicated on the graph representing the spatial distributions of randomly packed spheres up to 37% porosity which is the minimum possible for random close packed spheres.

generated sphere models and a naturally generated model for random close packing (Jerram *et al.*, 1996). Vectors for common processes have been established, like mechanical compaction, grain overgrowth and better sorting.

This method has considerable advantages as it is simple to apply. However, the validity of the reference models for more realistic crystal shapes and variable crystal sizes has not been tested. Also, the method gives only a single parameter that does not describe cluster size. More elaborate methods will be described in the following section.

If many or all crystals touch in three dimensions then the resulting framework will have an important effect on the rigidity of the material. Hence there is much interest in determining the minimum amount of crystals that is necessary to make such a framework. Jerram *et al.* (2003) used nearest-neighbour diagrams to determine if crystals in a rock form a touching framework. This technique has the advantage that it can be applied to sections; however, more data are needed before the method can be verified fully.

6.3.3.2 Cluster analysis

The goal of cluster analysis is to quantify the size and characteristics of clusters of grains in a rock. Clustering can reflect packing processes, the distribution of original nucleation sites or other processes (Jerram & Cheadle, 2000). The first attempt to quantify clustering sizes and distributions used the nearest-neighbour distance technique (see Section 6.3.3.1; Jerram *et al.*, 1996). More recently two, more complicated, techniques have been developed: complete linkage hierarchical cluster analysis (CLHCA) and density linkage cluster analysis (DLCA). These methods are ideally applied to the materials in which the complete, 3-D position of each grain is known. Jerram and Cheadle (2000) applied these methods to sections and tested performance with comparative 2-D sections through known 3-D synthetic textures (SAS package – see Appendix).

CLHCA was found to give the clearest results. It uses the distance between the grain centres. These are agglomerated to give the most compact groups. The result is a dendrogram (tree diagram) that links the grains and groups with lines. The distance between groups gives the 'height' in the tree. However, it is difficult to get direct quantitative information from a dendrogram – any distribution will give a dendrogram, including a random distribution. Another way of presenting the data is a graph of cluster frequency (number of clusters) against cluster size (number of grains in the cluster). At a cluster number of one all grains are included; at a cluster size of one grain the number of clusters equals the number of grains. This is a smooth curve for a random distribution, but has gentle steps for a clustered distribution. A further refinement can be made by normalising the data to a random distribution; deviations are then easier to see. This can be done by calculating the distance between clusters and subtracting the expected distance for a random distribution. A graph of this residual distance against cluster frequency gives the characteristics of the dominant clusters (Figure 6.4).

DLCA uses a different approach: clusters are defined on the basis of the amount of material within a certain radius. Hence this method considers the whole volume of the grain, and not just its centre. Clearly, this is a better approach if grains have a high concentration and many are touching. The technique can be applied easily to images with pixels or voxels. In many studies we can only observe grains in section, hence we are faced with the problem of calculating a 3-D parameter from 2-D data. Jerram and Cheadle (2000) tested this problem with a synthetic 3-D distribution, as before. They did not achieve good results, but this may reflect the weakness of the test distribution and not the method.

6.3.3.3 *Touching crystals*

Analytical methods discussed so far are only concerned with the position of the grain centre. They are used to reveal processes that acted early on grains. Conversely, the distance between the edges of the grains will reflect later

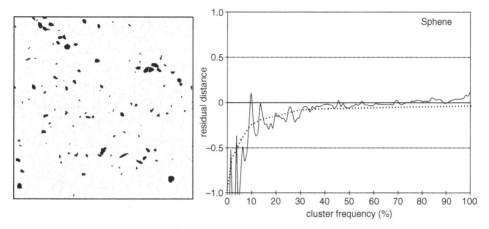

Figure 6.4   Complete linkage cluster analysis of sphene in a granulite sample (Jerram & Cheadle, 2000). Here the distance between the crystal centres is compared to that expected for a random distribution (residual distance). The dotted line is the random variation tolerance limit. Sphene crystals are more closely spaced than expected in the frequency 5–15%, which corresponds to cluster size of 1.5 to 5.1 mm.

processes. Obviously this can only be investigated completely in three dimensions. Unfortunately, with most 3-D analytical methods it is difficult to separate touching crystals and hence there has been no work on this subject. If the surface of contact between two grains is not parallel to the section plane, as is usually the case, then some, but not all, touching grains will be identified in sections.

Clusters of touching grains are frequently observed in volcanic rocks. In some cases the touching grains are sub-parallel (synneusis). Although it is possible to quantify this effect by looking at grain edge distance against differences in orientation, this has yet to be done.

Some rocks contain a touching network of plagioclase crystals that are visible in sections as a collection of crystal chains (Philpotts *et al.*, 1998). In three dimensions these chains must be projections of foam-like surfaces of crystals, but it is easier to continue the use of the term chain. The crystal chains are revealed most clearly when a block of basalt is melted: the block retains its shape even when 70% is liquid. Philpotts *et al.* (1998) identified the chains manually from reconstructed serial sections and X-ray tomography. There are many aspects of the chains that can be quantified, but so far the most useful has been the number density of the chains (Philpotts *et al.*, 1999). This was determined by counting the number of chains intersected along various random straight lines in sections. The number density can be transformed easily into the mean chain spacing, also called the network mesh size. The orientation of such chains is discussed in Section 5.8.3.1.

6.3.3.4 Wavelet transform and Fourier analysis

The wavelet transform is a powerful method to identify ordered structures in rocks (Gaillot *et al.*, 1997, Gaillot *et al.*, 1999). It can be used to identify the orientation and spacing of ordered structures and has been discussed in the section on grain orientation (see Section 5.4.3.3). Fourier analysis of grain sections has been used to look at grain distribution in sandstones (Prince *et al.*, 1995). Both these methods examine an image of the distribution of a mineral or grains, not individual grains.

6.3.4 Extraction of mineral association parameters

6.3.4.1 Mineral point correlations

Morishita and Obata (1995) proposed a rather different way of looking at the spatial distribution of grains and the relationship between grains of different minerals. It is based on the frequency of occurrence of pairs of points in a section with the same or different minerals versus distance. It uses a classified mineral map, but it is not necessary to define individual mineral grains. The method can only be applied where there is more than one phase in the rock. The method could also be applied in three dimensions.

  We will consider first a homogeneous, bimineralic rock, for simplicity. If two points are randomly selected then the *unconditional* probability that both are mineral 1 is V_1^2, where V_1 is the volumetric abundance of phase 1. Similarly, the probability that both are mineral 2 is V_2^2. The probability that we have chosen a mixed pair is $2V_1V_2$. We next consider the *conditional* probability of occurrence of mineral pairs at a specified distance D. These depend on the grain size and spatial distribution of the grains and are denoted $P_{11}(D)$ for a pair of points of mineral 1, $P_{12}(D)$ for a pair of points of mineral 1 and mineral 2, etc. For $D=0$ the points are coincident and the probability is equal to the volumetric abundance: $P_{11}(D)=V_1$. Similarly the probably that the points are different is zero: $P_{12}(D)=0$. Where D is very large the probabilities become equal to the unconditional probability, that is $P_{11}(D)=V_1^2$ and $P_{12}(D)=2V_1V_2$. The difference between the unconditional and conditional probabilities expresses the grain size, degree of order and degree of association of the mineral pairs. It is useful to define textural parameters that express these differences and extend the treatment to more than two minerals. For each mineral i a parameter $\sigma_i(D)$ is defined for a distance D:

$$\sigma_i(D) = \frac{P_{ii}(D) - V_i^2}{V_i - V_i^2}$$

For each mineral pair i, j a parameter $\tau_{ij}(D)$ is defined for a distance D:

$$\tau_{ij}(D) = \frac{P_{ij}(D)}{2V_i V_j}$$

At $D = 0$ all $\sigma = 1$ and $\tau = 0$; at $D = \infty$ all $\sigma = 0$ and $\tau = 1$.

Morishita and Obata (1995) examined a granite and a hornfels using these methods. They found that σ values tend to decrease exponentially with distance and the slope is a measure of mean crystal size. If the rock was ordered strongly then the σ values should peak at that distance, but this behaviour has not been recorded. They also found that τ values tend to increase monotonically with distance. Some mineral pairs rise to a peak and then go down, expressing a spatial association of those two minerals and giving a typical mineral spacing.

Morishita (1998) developed these measurements further by comparing σ and τ values of an 'ideal rock texture'. In this all crystals are randomly ($\equiv$ uniformly) distributed in the rock. He showed that if a number of assumptions are made then the crystal size distribution can be calculated from the variation in the probability that a point lies within a crystal against the distance from the crystal centre. However, it is not clear if this technique has any practical use, as it is not obvious how close natural rocks lie to this ideal texture.

This technique could certainly give interesting information on mineral associations, data which are not given by other methods. However, the lack of a readily available computer program to calculate the textural parameters has hampered the application of the method.

6.3.4.2 Box-counting method

Mineral associations can also be detected using a box-counting method (Saltzer *et al.*, 2001). This starts with a classified mineral image of a section of a rock. The image is first divided into four square areas each with an area of 2^{2n} pixels where n is the largest integer possible allowed by the image size. The value of n is decreased to zero, where each box contains one pixel. For each value of n the numbers of boxes dominated by two minerals are tabulated for each mineral pair. This is compared with a synthetically derived random distribution. A significant difference in the distributions indicates that the mineral pair is associated. This method does not quantify the correlation, but just indicates significant associations. The method does not seem to have any advantages over the method of Morishita and Obata (1995) and suffers from the same lack of an easy tool for doing the measurements.

6.4 Typical applications

One of the problems with this subject is that there is a large amount of theoretical and modelling work that produces patterns that resemble rocks (e.g. Ortoleva, 1994, Jamtveit & Meakin, 1999), but few actual measurements of the rocks themselves. Therefore, commonly, it has not been possible to verify if the proposed pattern-making processes actually occurred. In addition, many of the techniques presented in papers have not been taken up by other workers; hence the only example is the original paper. In some cases this is clearly a result of a lack of an easy-to-use computer program. A clear exception to this trend is the '*R*' parameter of Jerram *et al.* (1996).

6.4.1 Metamorphic rocks

Kretz (1966a) examined both the size distribution and position of garnet and biotite in a schist, and phlogopite, pyroxene and titanite in a marble. He measured the 3-D position of the crystals by disintegrating the samples with a small chisel. He found that the spatial distribution of garnet crystals was random and their size did not correlate with the distance to their nearest neighbour. Three years later he examined a section of pyroxene–scapolite–titanite granulite in more detail (Kretz, 1969). He divided up the section into sub-areas and measured the number density of crystals in each sector. By comparing the data to the expected abundance he was able to use a chi-squared test to show that the crystals were distributed randomly. He also showed that titanite was not associated preferentially with any of the major phases. He concluded that nucleation of all phases was random. The same section was re-analysed by Jerram and Cheadle (2000) using their cluster-analysis method. They found that the larger crystals of scapolite and pyroxene act as 'domains of order' – that is, there are fewer small nearest-neighbour distances than expected (see Figure 6.4). This method also showed that titanite is clustered.

6.4.2 Igneous rocks

Jerram *et al.* (2003) have reviewed data on clustering using a graph of their *R* parameter against per cent porosity (Figure 6.5). They find that crystals from a rhyolite laccolith are distributed randomly (see below; Mock *et al.*, 2003), but that all other examples are clustered. They established a field on this graph for touching frameworks by the inspection of samples. A good number of clustered examples with known touching frameworks have been used to define this region, but few non touching frameworks have been analysed, so perhaps

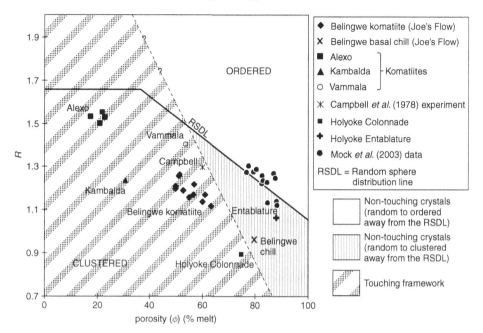

Figure 6.5   Porosity (%) or percentage melt against *R* value (from Jerram *et al.*, 2003).

more data are needed to establish this field more securely. Ikeda *et al.* (2002) examined two samples of different granites using the method of Jerram *et al.* (1996) and concluded that all minerals were clustered, except for biotite in one sample.

Boorman *et al.* (2004) examined the positions of orthopyroxene and plagioclase crystals in mafic rocks from the lower part of the Bushveld complex using the nearest-neighbour approach of Jerram *et al.* (1996). The *R* against porosity diagram was developed for spheres, but despite this limitation Boorman *et al.* (2004) considered that it could be applied to tabular crystals, such as those measured here (Figure 6.6). The vector defined by their samples could be the result of sorting, but does not correlate with the aspect ratio of the grains or the slope of the CSD, as would be expected. Instead, Boorman *et al.* (2004) suggest that this vector represents deformational compaction, similar to that proposed by Meurer and Boudreau (1998).

The spatial distribution of plagioclase, orthoclase and quartz in a rhyolite laccolith were examined by Mock *et al.* (2003). The more tabular feldspars seemed to show greater variation than the more equant quartz (Figure 6.5). There was a trend to clustering towards the top of the laccolith, and internal pulses of magma were distinguished.

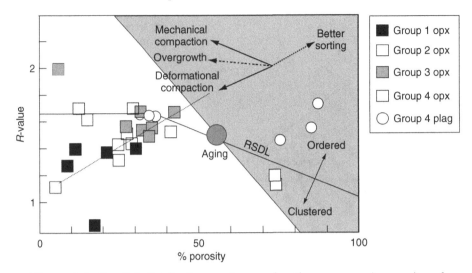

Figure 6.6 Spatial distribution patterns of orthopyroxene (squares) and plagioclase (circles) in the lower part of the Bushveld complex (Boorman *et al.*, 2004). Shaded region is for touching frameworks.

The more complex cluster analysis method of Jerram and Cheadle (2000) was applied by them to two samples. They showed that in a komatiite sample olivine crystals are clustered with diameters of 0.3–2.5 mm. In the scapolite–pyroxene–titanite rock examined by Kretz (1969) they found that titanite was clustered on a scale of 1.8–5.1 mm.

Chains of plagioclase crystals appear to be an important petrological structure in the thick Holyoke basalt lava flow and other slowly cooled bodies of similar composition (Philpotts *et al.*, 1999). The chains are a few crystals wide and consist of crystals up to 0.5 mm long attached in random orientations. They have been traced by hand from thin sections (Figure 6.7). They have been quantified by counting the number of intercepts per unit length and appear to be most numerous at the base of the flow and diminish upwards. The chains appear to originate in the upper part of the flow as groups of plagioclase crystals embedded in and surrounding pyroxene oikocrysts (Philpotts & Dickson, 2000). These crystal ensembles descend by convection to the lower part of the flow where the pyroxene exsolves and recrystallises into masses of granular pyroxene surrounded by the plagioclase chains. An investigation of the chains with a view to determining anisotropy produced by compaction is discussed in Section 5.8.3.1 (Grey *et al.*, 2003).

Igneous layering is commonly observed and has been the subject of numerous theoretical and qualitative studies (Parsons, 1987, Naslund & McBirney, 1996). However, so far there has been little quantitative work on crystal

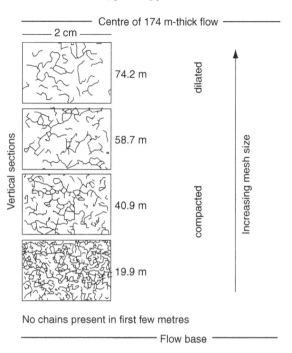

Figure 6.7   Plagioclase crystal chains in the Holyoke basalt flow (Philpotts *et al.*, 1999). Chains were traced by hand from thin sections.

positions in layered plutonic rocks. A good example is the inch-scale layering in the Stillwater intrusion. Boudreau (1987, 1995) has done a fine job of modelling the origin of the layering by self-organisation, but there are no quantitative studies of the grain positions and sizes in the rock.

There have been few studies of mineral associations: all are by the authors that originated the method and all just indicate significant associations without giving a value to this parameter. The spatial association of garnet and clinopyroxene in a sample of mantle peridotites was considered to be produced by mineral breakdown: orthopyroxene reacted to form garnet and clinopyroxene (Saltzer *et al.*, 2001). This was used to reconstruct a pressure–temperature history for these samples. Morishita and Obata (1995) examined two samples of hornfels and granite and found that biotite and quartz were spatially associated in the granite.

7

Textures of fluid-filled pores

7.1 Introduction

Many igneous and metamorphic rocks contain a fluid component or evidence of the former existence of such a fluid.[1] Fresh lavas and pumice blocks commonly contain vesicles, formed by the exsolution of gas from the silicate fluid. Metamorphic rocks may preserve evidence of partial melt pockets. The term 'pore' is used here to indicate space filled, or formerly filled, with fluid surrounded by a more viscous fluid (e.g. silicate melt, glass) or crystals. Pores may be considered as the fluid equivalent of grains or crystals in solid phases. Fluid inclusions in crystals are excluded here as their textures have yet to be studied quantitatively.

Pores do not always stay filled with their original fluid during lithification. For instance, vesicles may be filled with other minerals to become amygdules. In partially molten silicate rocks the fluid may quench to a glass, or a mixture of minerals, whose textures pseudomorph the original pores. Hence the porosity of a rock, as discussed here, is the original volume proportion of fluid at the time of interest. For example, the porosity of a lava sample is the volumetric proportion of vesicles at the moment of solidification; the porosity of a partially molten rock is the volumetric proportion of fluid when the texture is preserved and such a melt may have gas bubbles. Clearly, it is important to specify what feature is being examined. The porosity is characterised at a first approximation by the total volumetric proportion of fluid, but like grain size, it is also possible to quantify the size distribution of pores.

Surfaces are important in many geological processes, as sites of chemical or mechanical processes. The pores define an internal surface area in a rock. The outside of grains defines the external surface area, which may be significant for loose materials. Clearly some materials, such as pumiceous pyroclastic deposits will have significant internal and external contributions to the total surface area. In natural materials surfaces are complex, hence the actual value of the total surface area depends partly on the scale of the analysis and the measurement method (Brantley *et al.*, 1999, Brantley & Mellott, 2000).

Another important associated property is the permeability of the rock. In many ways it is much more important than the porosity, as this parameter measures how readily a fluid can flow through a rock. The total permeability can be measured, in addition to the size distribution of pore channels and 'throats' that control fluid flow. Rocks with a high porosity tend to be highly permeable, but the porosity–permeability relationships diverge as the porosity goes down.

7.2 Brief review of theory

7.2.1 Permeability

The permeability of a rock is used qualitatively to describe the ability to conduct fluid: a more permeable rock conducts fluid more easily that a less permeable rock. Hence the permeability expresses the connectedness of the pores. The original development of the subject considered the flow of water in a rock and is expressed as Darcy's law:

$$\Delta P/l = \nu\mu/k$$

where ΔP is the pressure drop over a length l, ν is the specific discharge or filter velocity, μ is the dynamic viscosity of the fluid and k is the permeability. The permeability is sometimes expressed in Darcy units, which are equal to about $10^{-8}\,cm^2$. The permeability of a rock is not necessarily the same in all directions.

Darcy's law is only applicable in low Reynolds number (ratio of inertial forces to viscous forces) flow where energy loss is due to viscous effects. At higher flow rates the fluid inertia becomes important and another term must be added to Darcy's law (see review in Rust & Cashman, 2004):

$$\Delta P/l = \nu\mu/k_1 + \nu^2\rho/k_2$$

where k_1 is the viscous (Darcian) permeability, ρ is the fluid density and k_2 is the inertial (non-Darcian) permeability.

7.2.2 Magmas and partially molten rocks

Movement of fluid with respect to solids is a fundamental process in the solidification of igneous rocks and the melting of metamorphic rocks. It may occur by compaction (McKenzie, 1984), porous-media convection or in response to shear. It is especially important in the production of monomineralic rocks, such as anorthosite or dunite. The final porosity is commonly expressed as the 'trapped liquid' component and can be determined by chemical or textural methods. The key property is clearly the variation in permeability during solidification or melting: if the permeability is finite for all values of porosity, then all the silicate fluid can leave the rock. However, it is commonly proposed that there is a percolation threshold, below which the permeability falls to zero. The actual value of such a threshold may be partly controlled by the interfacial angles of the fluid with the solid phases, which in turn express differences in interfacial energy (see Section 4.2.3).

Cheadle *et al.* (2004) have reviewed and investigated the relationship between textural equilibrium, dihedral angles and permeability using a geometrically simple numerical model. The simplest models are for settled or random rigid shapes. These are termed non equilibrated as their shape is unchanged after accumulation. For these materials the percolation threshold varied with crystal shape – it was about 3% for spheres and 10% for cubes (Figure 7.1). This means that there will always be a minimum of 3 to 10% melt trapped in the cumulate. For texturally equilibrated samples, with dihedral angles less than 40°, the percolation threshold was close to zero. That is, all interstitial fluid (silicate liquid) can escape and monomineralic rocks can form.

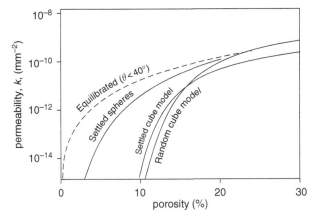

Figure 7.1 Melt permeability against porosity for a crystal–liquid system. For the settled and random models the porosity is the space between crystals represented as simple geometrical forms. The crystals are allowed to interact in the equilibrated model, giving dihedral angles, θ, of less than 40°. From Cheadle *et al.* (2004).

7.2.3 Bubbles in magmas

7.2.3.1 Nucleation and growth of bubbles

Bubbles nucleate in a similar way to crystals (Figures 7.2a, 7.2b). Nucleation occurs when the gas becomes supersaturated in the magma. Classical nucleation theory (e.g. Gardner & Denis, 2004) predicts that the activation energy, ΔF, for formation of a bubble nucleus in a homogeneous fluid is

$$\Delta F = \frac{16\pi\sigma^3}{3\Delta P^2}$$

where σ is the surface tension of the bubble, and ΔP is the supersaturation pressure.

This equation must be modified for heterogeneous nucleation (on a crystal surface): σ becomes the balance of surface tensions for crystal, melt and bubble and the activation energy is modified by a factor ϕ which is related to the wetting angle (solid–liquid dihedral angle; see Section 4.2.3) of the bubble on the surface (θ; Figure 7.3): $\phi = 1$ for completely wetted surfaces (identical to homogeneous nucleation) and $\phi = 0$ for a wetting angle of $180°$ (Gardner & Denis, 2004). Application of this simple theory predicts that homogeneous nucleation requires a high degree of supersaturation and hence may be rare. However, heterogeneous nucleation requires a much lower degree of supersaturation, and hence will dominate if crystals are present. For example, iron oxide crystals appear to be better nucleation sites as bubbles do not wet the surface, whereas plagioclase is wetted by bubbles and hence requires a very high supersaturation pressure (Figure 7.3; Hurwitz & Navon, 1994, Gardner & Denis, 2004).

After nucleation bubbles grow by diffusion of vapour through the liquid (Figure 7.2c). The speed of this process will depend on the viscosity of the liquid and the molecular weight of the vapour: diffusion is fastest for light molecules in a low-viscosity fluid. There is a feedback between bubble growth and nucleation of new bubbles. Growth of bubbles by diffusion reduces the vapour content of the melt, which will in turn affect the surface tension. A lower water content in the melt will require a higher supersaturation pressure; hence nucleation of new bubbles will be more difficult. The overall result of this feedback is that nucleation may be instantaneous, rather than continuous.

Once bubbles have formed they will increase in size when the magma moves to a lower pressure region (Figure 7.2d). Expansion will be slowed down by viscous effects and relatively large changes in depth are probably needed to make significant changes in bubble size.

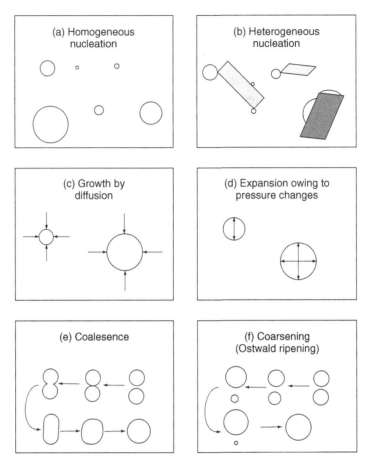

Figure 7.2 Bubble formation, growth and shrinking. (a) Homogeneous nucleation occurs within a fluid. (b) Heterogeneous nucleation occurs on a surface. If the angle between the bubble and the surface is small the surface is said to be wetted and nucleation requires a higher degree of supersaturation. (c) Growth of pores can occur by diffusion of fluid into the pore. (d) If the overall pressure is reduced bubbles will expand provided the matrix has a low enough viscosity. (e) Bubbles may also grow by coalescence. (f) Small bubbles have a higher internal pressure than larger bubbles. As the material approaches equilibrium vapour will be transferred via the matrix, leading to a coarsening of the bubbles.

The kinetically controlled processes of bubble nucleation and growth can continue as long as there is sufficient supersaturation. As the driving force for these processes diminishes, bubbles will try to approach equilibrium by size adjustments. Bubble coalescence occurs when two bubbles touch and become one: it has no parallel in crystal growth (Figure 7.2e). Rates of coalescence depend on the timescale of bubble approach, film thinning, rupture and

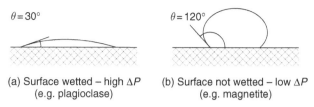

(a) Surface wetted – high ΔP
(e.g. plagioclase)

(b) Surface not wetted – low ΔP
(e.g. magnetite)

Figure 7.3   Nucleation bubbles on surfaces. (a) Wetted surfaces have low dihedral angles ($\theta = 30°$) and require large supersaturation for nucleation. (b) Unwetted surfaces ($\theta = 120°$) require lower degrees of supersaturation.

relaxation. Experimental results suggest that coalescence is most important when the bubbles are expanding in response to a reduction in the confining pressure (Figure 7.2d; Larsen *et al.*, 2004). Gaonac'h *et al.* (1996a) have proposed that cascading bubble coalescence dominates other equilibrium processes. During the final stages of development cascading bubble coalescence can lead to catastrophic fragmentation of the magma (Gaonac'h *et al.*, 2003).

Coarsening (Ostwald ripening) of bubbles resembles the same process in crystals (Figure 7.2f; see Section 3.2.4). A larger bubble has a lower internal pressure than a smaller bubble. Hence, gas will move from the smaller bubble to the larger by diffusion through the silicate liquid. Clearly, the importance of this process depends on the rate of diffusion of gas through the silicate liquid, which in turn is controlled by the distance, temperature, liquid viscosity and gas composition. However, unlike coalescence, it will occur even if the bubbles do not intersect (Larsen & Gardner, 2000).

7.2.3.2 Bubble size distributions and shapes

Bubble size distributions (BSDs) can follow many different models, such as unimodal, polymodal, exponential and power law (Blower *et al.*, 2002, Gaonac'h *et al.*, 2003). Unimodal BSDs are produced by instantaneous bubble nucleation, most likely following a heterogeneous model (Blower *et al.*, 2002). Polymodal BSDs probably form by several instantaneous nucleation events.

Both exponential and power law BSDs are commonly observed. The similarity of exponential BSDs to some CSDs (Marsh, 1988b) has suggested to some authors that they can form in the same way, by continuous nucleation and growth under steady-state conditions (e.g. Mangan & Cashman, 1996). Power law BSDs may form by cascading coalescence of bubbles (Gaonac'h *et al.*, 2003). However, other models are possible: Blower *et al.* (2002) have proposed a model of non-equilibrium continuous nucleation and growth of bubbles that evolves from a unimodal distribution via an exponential model to a power law distribution.

The shape of bubbles in magma can be changed by deformation (Rust *et al.*, 2003). These shapes will be preserved if the relaxation times of the magma are long, as is the case for a viscous magma such as rhyolite. However, non-spherical bubbles are not usually preserved in mafic magmas, except near the chilled surface. The shape of bubbles has been used to quantify the type of shear, simple or pure, and the shear rate (Rust *et al.*, 2003).

7.3 Methodology

The overall porosity and pore size distribution of a material can be measured using 3-D and 2-D methods, to be discussed below. Pores are commonly assumed to approximate to spheres, hence their size is considered to be the diameter of a sphere of equivalent volume (Figure 7.4). If the pores are connected then the size of the pore is calculated assuming that the pore is closed at the pore throat. For some strongly anisotropic materials pores may depart from such a model and then size definitions similar to those used for grains may be used (see Section 3.2.1.1). For some applications the actual volume of the pore is used instead of the diameter.

The permeability of a material depends on the connections between the pores (Figure 7.4). Connections may be increased by fracturing of the material

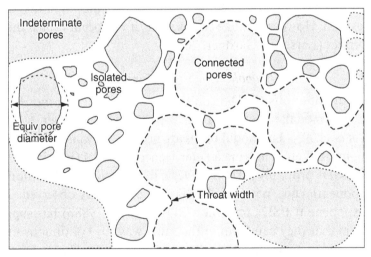

Figure 7.4   Schematic 2-D section of a porous material. Pores were or are filled with fluids, such as oil, gas, water or silicate melts. Pores may be divided into those that are isolated (solid edges) and those connected by throats (dashed edges). This is a section so isolated pores may be connected in three dimensions outside the plane of the section. It is not possible to classify pores that intersect the edge of the sample (dotted edges).

during preparation. If the sample is small then permeability may be dominated by edge effects: all pores that intersect the surface of the sample are connected to it. However, their connections in the original rock are unknown. Hence, a researcher should be aware of connections between the size of the sample and the resolution of the method.

7.3.1 Pore surface area measurements

7.3.1.1 Analytical methods

Nitrogen adsorption is a physical method for measuring the total surface area of a material permeable to gas (Gregg & Sing, 1982, Whitham & Sparks, 1986). It cannot measure the area of isolated pores. The sample is first dried in a vacuum. Then a stream of nitrogen is passed over the sample and allowed to come to equilibrium. The nitrogen is adsorbed as a single layer of molecules on the surfaces. Pure helium is then passed over the sample and the nitrogen releases from the surface. The amount of nitrogen in the helium is measured. The experiment is repeated for different partial pressures of nitrogen. From these data the surface area per unit volume can be determined. If a shape for the pores is assumed, such as a cylinder, then the mean pore diameter can be calculated.

Mercury intrusion porosimetry involves the adsorption of mercury by a permeable rock under increasing pressure (Whitham & Sparks, 1986). The adsorption is opposed by the surface tension of the mercury. This in turn depends on the wetting angle and the diameter of the pores. The geometry of the pores must be assumed – a cylinder is the most common shape used. The measurements are done in two apparatuses: the sample is first put in a vacuum and mercury adsorbed, typically at a pressure of 0.078 bar, which corresponds to a pore diameter of 235 µm. The pressure is increased to atmospheric, equivalent to a diameter of 15 µm. Finally the sample is transferred to a pressure vessel where the pressure is gradually increased: a pressure of 4280 bar corresponds to a pore diameter of 0.0032 µm. This method is useful for the smaller pores, but again only counts pores connected to the outside of the sample. However, it can be expected that thin pore walls will fracture under these high pressures. Total porosity can also be determined in a similar way by adsorption of water under pressure (Whitham & Sparks, 1986).

7.3.1.2 Image analysis method

The surface area per unit volume is one parameter that can be determined exactly from intersection data (see Section 2.7.2). Although this method is independent of pore shape, connectedness or orientation, it is best to use a number of orthogonal sections for markedly anisotropic materials. The sections are imaged using optical or electronic methods (see Section 2.2). Test lines are drawn on the images and the number of intersections of the line and the pores counted. The surface area per unit volume is equal to twice the number of intersections divided by the length of the test line. This method counts all pores, including the isolated pores missed by the nitrogen adsorption method.

7.3.2 Three-dimensional analytical methods

7.3.2.1 Total porosity

Total porosity can be calculated easily from the density of the rock, that of the void-free rock and the void-filling material. The density of the rock is most easily determined by cutting an orthogonal block and measuring and weighing it. Archimedes' principle can also be used provided that the measuring fluid, usually water, does not get into the pores (Houghton & Wilson, 1989). If the pores are less than 1 mm then the block can be sprayed with waterproofing aerosol as used on camping gear. For larger pores the block can be wrapped in wax sheet, whose mass must be considered. Rocks which are less dense than water can be ballasted with a known weight. For rocks with a known mineralogical composition the void-free density can be calculated by summing the proportion of the minerals weighted for their density. For igneous rocks the density can be estimated from the chemical composition (Ghiorso & Sack, 1995). This parameter is calculated by a number of geochemical programs (see Appendix).

The true rock density can also be measured by helium pycnometry: the volume of helium displaced by the sample increases the pressure in a chamber and the density can be calculated using the ideal gas law (Klug & Cashman, 1994). Helium will only enter into all pores connected to the surface hence the total porosity may be less than the density method described above. The measurements are done in a dedicated instrument.

7.3.2.2 Three-dimensional imaging techniques

The distribution of pores can be determined by serial sectioning (see Section 2.2.1). In transparent materials, such as volcanic glasses, the size of bubbles can be measured directly with a regular or confocal microscope (see Section 2.2.2).

X-ray tomography methods similar to those used for measuring the distribution of crystals can be used for pores also (see Section 2.2.3). Synchrotron

X-ray microtomography is very similar, but uses a more intense source with a finer resolution (Song *et al.*, 2001, Shin *et al.*, 2005). Finally, magnetic resonance imaging has been used to visualise the texture of carbonate sediments (Gingras *et al.*, 2002). This method is particularly sensitive to water distribution. However, the resolution of the method is poor and there are few advantages over other methods.

7.3.3 Two-dimensional analytical methods

7.3.3.1 Measurement of pore outlines

The most popular method for determining porosity and pore size distributions is image analysis of sections or surfaces. The open spaces in rocks may be made more visible in thin section if the rock is first impregnated under vacuum with coloured epoxy resin. This process clarifies the distinction between isotropic materials like volcanic glass and the open pores. In slabs it is better to use a fluorescent dye. With this method pores down to 1 μm can be imaged. Some workers have used low-melting-point metals for impregnation, such as Woods metal. Unfortunately this alloy is rich in cadmium and hence toxic.

Sections can be imaged in transmitted or reflected light (see Section 2.5). They can also be examined using a scanning electron microscope. Images can be processed and segmented using the same methods as for minerals (see Section 2.3).

7.3.3.2 Conversion of two-dimensional data to 3-D pore parameters

Sectional data must be converted to 3-D data using stereologically correct methods, such as those developed for grain and crystal intersections (see Section 3.3.3). The spherical or sub-spherical shape of many pores, especially in volcanic rocks, makes correction for the cut-section effect less significant than for more complex shapes. However, the intersection-probability effect remains just as strong.

Small pores (<30 μm) may be seen in projection when examined in thin section. In this case the equations for conversion of data are different from those used for intersection (see Section 3.3.4). For spherical pores the data do not need any correction at all. Gardner *et al.* (1999) used a variant of the projection method: they counted the number and size of bubbles within a small volume and then used the total bubble volume to scale the measured bubble size distribution to the values for unit volume.

Thick sections with smaller pores viewed in projection and larger pores viewed in intersection are difficult to treat. Ideally the two populations must be separated, calculated separately using the appropriate methods and then recombined.

Blower *et al.* (2002) have proposed a parametric model for conversion of sectional data. They note that if the pores are spherical and their sizes follow a power law distribution, then the distribution of sectional diameters will also have a power law distribution. Hence, the parameters of the 3-D power law can be determined from the 2-D power law. Parametric solutions are always dangerous to use as pore and grain sizes rarely conform to predetermined distribution laws. In fact, we are generally searching for the distribution and hence it is unwise to assume it beforehand (see Section 3.3.4).

Some early studies used the 'Wager' equation to convert intersection size data to pore size distributions (e.g. Mangan *et al.*, 1993). As was discussed in Section 3.3.4.6 this equation is not appropriate, and does not give correct results, but published results can commonly be recalculated using more accurate methods (Higgins, 2000).

Toramaru (1990) proposed the use of the 'Spektor' method for determining the pore size distribution. This method gives correct results, but necessitates measurement of a very large number of pores to give precise results. It is usually better to measure pore intersection size directly and make stereological corrections.

Pore shapes and orientations can be determined using the same methods as for solid grains (see Sections 4.2 and 5.2).

7.3.4 Measurement of permeability

Permeability is generally measured in a dedicated instrument. For example, for a volcanic rock the sample is a cylinder of rock, typically 25 mm in diameter and 23 mm long (Rust & Cashman, 2004). The edge of the core is sealed with a rubber sleeve, leaving the circular ends open. Gas is forced through the sample. The pressure difference between the two ends is increased in many small steps from zero to 1.4 bar and the gas flow rate measured. From this the permeability parameters for the viscous and inertial components of the permeability can be determined.

7.4 Parameter values and display

7.4.1 Display of pore size distributions

Pore size distributions (PSD) can be expressed and displayed in a number of ways. Some of these are similar to those used for grain size distributions (see Section 3.3.7). As for grain size distribution, it is always important to recognise the resolution limits of any analytical method. These will clearly affect the interpretation of the pore size distribution and the values of parameters that describe the pores.

A diagram similar to that used for crystal size distributions can be used (see Section 3.3.7.2; Marsh, 1988b). In this graph the natural logarithm of the population density is plotted against size, generally expressed as the equivalent radius or diameter. As before population density must be distinguished clearly from volume number density (see Section 3.3.7.2). This diagram is useful if the pore population is generated by nucleation and growth in a similar way to crystals. However, in many situations nucleation is instantaneous and hence the Marsh model is not applicable.

Many authors have used simple frequency histograms to display pore size distributions. The horizontal axis is either size as equivalent radius or volume expressed linearly or in log 2 units (Toramaru, 1990). However, as was mentioned in Section 3.3.7, this is not a very good diagram if the data are sparse, as they tend to be for the larger size classes. In this situation many classes will only contain one or two pores or may be empty (e.g. Larsen *et al.*, 2004). The eye tends to glide over the empty classes, creating the impression that the number of larger pores is greater than it is. This problem cannot be resolved by increasing the width of classes as then the height of the histogram bar will change. A better, and equally simple, diagram uses frequency density. The frequency in each size class is divided by the width of the class. Larger classes can then be wider, eliminating empty or sparse size classes that are just artefacts of measurement.

Other distribution models can be verified by suitable diagrams in the same way as for crystal size distributions (see Section 3.3.7). For example, Klug *et al.* (2002) used a cumulative frequency diagram to verify if BSDs were lognormal and a log–log diagram to determine if a power law (fractal) distribution was a better approximation.

The connectedness of pores can also be quantified. It can be expressed in terms of the number density (number per unit volume) of interconnected pores against pore volume (Song *et al.*, 2001). Pores are connected by pore throats and the dimensions of such throats can also be expressed by a mean value and size distribution.

7.4.2 Overall pore size parameters

The moments of the pore size distribution, M_0, M_1 etc. can be calculated easily and are simply related to the overall properties of the pores. These equations are slightly different from those in Section 3.3.4 as the pores are assumed to be spheres and the size variable is the radius of the pores.

M_0 = total number of pores per unit volume, N_V.
M_1 = total radius of pores = $N_V \times$ the mean radius of the pores, R_m.
M_2 = total projected area of pores = $1/4\pi \times$ surface area per unit volume, S_V. However, S_V can be obtained precisely by other methods that do not need to assume a pore shape (see Section 2.7.2).
M_3 = total volume of pores = $3/4\pi \times$ porosity. Again, the total porosity can be obtained precisely by other methods that do not need to assume pore shape (see Section 2.7.2).

7.4.3 Pore shapes

The shape of pores can be quantified in the same way as for crystals and grains (see Section 4.3.3). However, simple aspect ratio parameters are used most commonly. Small vesicles commonly have the form of an ellipsoid and hence their shape can be specified by their aspect ratio (e.g. Rust *et al.*, 2003). Pore orientation can again be quantified in the same way as for grains (see Section 5.5). A commonly used measure is the orientation of the long axis of the pore with respect to a structurally important direction, such as flow for lavas. The overall texture of pores can also be described in terms of fractals (Meng, 1996).

7.5 Typical applications

Pore size distributions (PSDs) have been measured in a wide variety of rocks. Some of the early studies that determined PSDs from measurements made on sections used incorrect conversion methods; hence the data must be reconverted before it can be compared with measurements made in three dimensions or using more accurate stereological techniques. The equivalent problem for crystals is discussed in Section 3.3.4.6.

Partially molten rock contains silicate-liquid-filled pores and can be studied in the same way as other porous materials. In some cases the liquid is preserved as a glass, but in many plutonic rocks its presence must be inferred from the textures of minerals considered to pseudomorph the original melt-filled pores. So far pore structure has been explored by measurement of dihedral angles of solid–solid–liquid intersections and these results are discussed in Chapter 5.

7.5.1 Volcanic rocks

Quantitative study of vesicles is a very active research field of volcanology because nucleation and growth of bubbles can trigger and drive volcanic eruptions (e.g. Sparks, 1978). Vesicle size distributions (VSDs) are studied because they may give information on degassing and magma volatile contents (Toramaru, 1989, Gaonac'h *et al.*, 1996b, Larsen & Gardner, 2000) and may be used to determine eruption conditions (Blower *et al.*, 2001).

Explosive silicic volcanic eruptions are not only important for their geological effects, but also for their potential impact on human communities. One of the larger Holocene eruptions was that of Mt Mazama, which produced Crater Lake (Klug *et al.*, 2002). The main product of this eruption was highly vesicular pumice in which most pores are connected. The high vesicle number densities indicate that bubble nucleation occurred rapidly at a high degree of supersaturation; hence continuous nucleation models cannot be applied. Small vesicles are approximately lognormal by volume reflecting this single nucleation and growth event. Larger vesicles follow a power law distribution which may be due to coalescence (Figure 7.5). Klug *et al.* (2002) proposed that nucleation was in response to a downward propagating decompression wave, followed by bubble growth and coalescence, and finally fragmentation.

Submarine pyroclastic volcanism seems at first glance to be unlikely, but there is abundant evidence of its occurrence in some situations. Clearly, the question involves degassing, hence the interest in evidence provided by VSDs

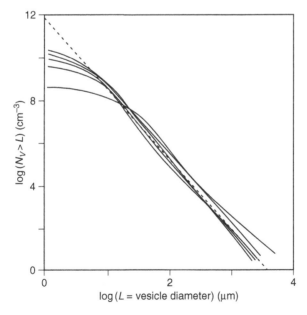

Figure 7.5 Cumulative vesicle size distributions in pumice from Mt Mazama (Klug *et al.*, 2002). Size distributions of all samples depart from a power law model (slope 3.3) at small sizes.

in pumice. The Healy caldera is an active volcano currently at a depth of 1.1–1.8 km. Wright *et al.* (2003) have quantified the shape and size of vesicles in pumice from this eruption. There is a good correlation between vesicle size and shape: increasing complexity with size is considered to reflect coalescence. VSDs were measured by integrating slab and thin section data. Unfortunately, the data were converted using inappropriate methods (see Section 3.3.4.6). Despite this the VSD largely follows a power law between 0.03 and 2 mm with $d = 2.5 \pm 0.4$ (the effect of the conversion error on the VSD is unclear). Departures from the distribution outside those limits are probably measurement artefacts. The VSD, magma water content and temperature indicate that pyroclastic eruptions could have occurred at depths of 500–1000 m. As the pyroclasts cooled they ingested water and sank.

The vesicle size distribution of lava flows has been proposed as a measure of palaeoelevation of an eruption (Sahagian & Maus, 1994). The basis of this method is that there is a difference in pressure between the base and top of the flow caused by the mass of the lava. This pressure can be calculated and used to calibrate the correlation between bubble sizes and pressure. The method can only be applied to flows that have not been inflated after the bubbles were frozen in at the base and top. The method was first applied to recent lavas from Mauna Loa, Hawaii, and found to have a precision of about ±400 m (Figure 7.6; Sahagian *et al.*, 2002a). It was subsequently applied to the problem

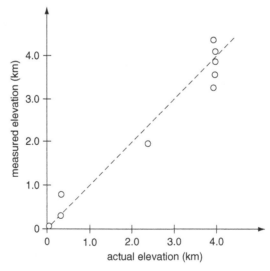

Figure 7.6 Elevation determinations from differences in vesicle sizes between the top and base of a flow against actual elevations (after Sahagian *et al.*, 2002a). These data are from recent lavas on the slopes of Mauna Loa, Hawaii.

of the timing of uplift of the Colorado plateau during the last 23 Ma (Sahagian *et al.*, 2002b). Bubble size distributions indicated that the uplift rate was low until 5 Ma ago when it increased by a factor of four.

Bubble shapes and orientations can also be used to infer flow type, shear rate and shear stress in magmas (Rust *et al.*, 2003). Bubbles in volcanic silicic glasses are almost ideal strain-rate markers as they can be measured easily in three dimensions and their relaxation times are long. The shape of bubbles is a balance between shear forces that deform bubbles and surface tension (interfacial energy) that rerounds bubbles. Bubble shapes in three obsidian flows were quantified in terms of bubble shape, size and orientation. These data immediately provide data on the flow type: a spatter-fed flow deformed by pure shear whereas simple shear dominated in juvenile obsidian from a pyroclastic deposit and a well-banded obsidian flow. Additional data on magma viscosity and surface tension enabled the calculation of shear rates and shear stresses.

Some glassy basaltic samples dredged from the ocean floor are so charged with gas that they pop when brought to the surface. Burnard (1999) examined one such sample that had 17% vesicles. He quantified the shape of the vesicles against size and concluded that degassing occurred whilst the magma was rising in a conduit at a speed of at least $1\,\mathrm{ms}^{-1}$. This is comparable with some explosive eruptions in Hawaii. Bubble size distributions were semilogarithmic, indicating that nucleation of bubbles was continuous and not a single event.

Degassing and crystallisation are clearly linked in many volcanic systems, but there have been few studies in which the two have been linked quantitatively. Gualda *et al.* (2004) have investigated quartz size distributions and total vesicularity in the rhyolitic Bishop Tuff. They found that there was an inverse correlation between the total vesicularity and the amount of the largest fraction of quartz crystals. They suggested that it was the result of the sinking of large crystals and the rise of bubbles in the magma chamber. However, it is also possible that the link is loss of gas during coarsening of the quartz in the magma. Very long magma storage times are certainly consistent with this idea.

Notes

1. Fluid is used here in its physical sense for a gas, liquid or supercritical fluid. It therefore includes fluids with a wide range in chemical composition, such as those dominated by water, hydrocarbons, carbon dioxide, silicates, carbonates, sulphides, etc.

8

Appendix: Computer programs for use in quantitative textural analysis (freeware, shareware and commercial)

An updated version of these pages will be available at http://geologie.uqac.ca/%7Emhiggins/CSD.html

8.1 Abbreviations

C++	Uncompiled program in C++
FORTRAN	Uncompiled program in FORTRAN
IDL	IDL language
Mac	Macintosh
MacX	Macintosh OS X
Win	Windows
Linux	Linux
All	All of the above platforms
Excel	Excel worksheets and macros
NIH	NIHImage (ImageJ) macros
Free	Freeware
Share	Shareware
Com	Commercial

8.2 Chapter 2: General analytical methods

This program can be used to segment, separate and measure data volumes (3-D images).

BLOB3-D	IDL, All	Ketcham (2005)
ftp://ctlab.geo.utexas.edu/blob3d		

These programs can be used to combine serial section images into a 3-D image.

| NIHImage
http://rsb.info.nih.gov/
  nih-image/ | Mac, Free | This is a very good freeware program. Many versions are available |
| ImageJ
http://rsb.info.nih.gov/ij/ | Win, Mac, MacX,
Linux, Free | This is a java version of NIHImage. It is being upgraded and maintained |

Bitmap images (jpeg, tiff, bmp, etc.) can be stitched together with these programs.

Adobe Photoshop http://www.adobe.com/	Win, Mac, Com	The basic version is bundled with some digital cameras
ArcSoft Panorama maker	Win, Mac, Com	Bundled with some digital cameras
PanaVue ImageAssembler http://www.panavue.com	Win, Com	Can stitch very large images

Images acquired by scanning photographs, rock blocks or thin sections must be converted to binary images before they can be analysed automatically. There are a number of commercial and shareware programs available including:

Adobe Photoshop http://www.adobe.com/	Win, Mac, Com	This is the classic image processing program
Corel PhotoPaint http://www.corel.com/	Win, Mac, Com	This program is very similar to Photoshop and comes included with Corel Draw
LViewPro http://www.lview.com/	Win, Share	This program is quite compact and does a lot for a shareware program
Image-Pro Plus http://www.mediacy.com/	Win, Com	This program is very good but expensive
IPLab http://www.iplab.com/sos/ product/gen/IPLab.html	Mac, Win, Com	Commercial image analysis program

Lazy grainboundaries
http://pages.unibas.ch/
  earth/micro/index.html/
  grainsize/grainsize.html
	NIH, Free	Automatic tracing of grain boundaries

Binary images must be analysed to extract textural information. These programs can do this.

Image-Pro Plus http://www.mediacy.com/	Win, Com	This program is very good but expensive
IPLab http://www.iplab.com/sos/ product/gen/IPLab.html	Mac, Win, Com	Commercial image analysis program
Aphelion http://www.adcis.net	Win, Linux	Commercial image analysis program
NIHImage http://rsb.info.nih.gov/ nih-image/	Mac, Free	This is a very good shareware program. Many versions are available
ImageJ http://rsb.info.nih.gov/ij/	Win, Mac, MacX, Linux, Free	This is a Java version of NIHImage. It is being upgraded and maintained
ImageTool http://ddsdx.uthscsa.edu/dig/ itdesc.html	Win, Free	Freeware image processing, originally developed for biological work
Visilog http://www.norpix.com/ visilog.htm	Win, Com	Commercial image analysis program

Digitising tablets and control programs.

Sigma-Scan http://www.spss.com/software/science/SigmaScan/	Win, Com	Made by Jandel Scientific.

These special programs treats multiple images acquired at different orientations of the polarisers with computer-integrated polarising microscopy (CIP).

Geovision http:// craton.geol.brocku.ca/ faculty/ff/stage/ WelcomeF.html	Win, Free	Petrographic image processing of thin sections using the rotating polariser stage.
Image segmentation ftp://ftp.iamg.org/VOL30/ v30-08-03.zip	Linux, Free	Segmentation of petrographic images acquired with CIP microscopy. (Zhou *et al.*, 2004b)

Miscellaneous programs and sites.

MELTS http://ctserver. uchicago.edu/	Mac, Unix, Java	Calculation of liquid line of descent, density and other parameters (Ghiorso & Sack, 1995)
MinIdent-Win http://www. micronex.ca/	Win, Com	Mineral data database: mineral properties include chemical composition, optical, X-ray, physical, hand-specimen. Data ranges compiled from real analyses. About 6000 named and unnamed minerals
International Society for Stereology http://www.stereology society.org/		

8.3 Chapter 3: Grain and crystal sizes

CSDCorrections http://geologie.uqac.ca/ %7Emhiggins/ csdcorrections.html	Win, Free	Distribution-free methods. Blocks, ellipsoids, massive and oriented textures. Log or linear size intervals. Choice of output graphs. Michael D Higgins, UQAC
CSD – Excel spreadsheet (contact author)	Win, Free	Parametric solutions – Tony Peterson, Geological Survey of Canada
StripStar http://pages.unibas.ch/earth/ micro/index.html	Mac, Free	Distribution-free methods. Conversion of spheres only. Only uses fixed with bins. Not user friendly. M Strip, University of Basel

8.4 Chapter 4: Grain shape

ImageJ http://rsb.info. nih.gov/ij/	Win, Mac, MacX, Linux, Free	This is a Java version of NIHImage Box-method fractal parameters can be determined
Fractscript (contact author)	ImageTool macro	A macro for calculating the fractal dimension of object perimeters in images of multiple objects (Dillon *et al.*, 2001)
Fraccalc (contact author)	NIH	M. Warfel in Dellino & Liotino (2002)

Gaussian Diffusion-Resample (contact author)	NIH	(Saiki, 1997, Saiki *et al.*, 2003)

8.5 Chapter 5: Grain orientations (rock fabric)

INTERCPT and SPO http://www.sciences. univ-nantes.fr/geol/ UMR6112/SPO/ index.html	Win, Free	Program for applying the inertia tensor and intercept method to binary bitmap images. Patrick Launeau and Pierre Robin
Ellipsoid http://www.sciences. univ-nantes.fr/geol/ UMR6112/SPO/ index.html	Win, Free	Ellipsoid calculation of 3-D shape preferred orientation by combination of 2-D images. Patrick Launeau and Pierre Robin
Various programs http://www.sp.uconn.edu/ ~geo102vc/Compaction_ Programs.htm	Win, Excel, Mathlab, Free	Programs to estimate compaction from scanned thin section images and digitised plagioclase chain networks. N. H. Gray and A. Philpotts
Basel software package http://www.unibas.ch/ earth/micro/software/	Mac, NIH, FORTRAN	Package of programs and macros for computer-integrated polarisation microscopy (CIP)
Lazy macros http://www.unibas.ch/ earth/micro/software/ macros/lazymacros.html	NIH, Free	Package of programs and macros for computer-integrated polarisation microscopy (CIP)
Wavecalc http://www.iamg.org/ CGEditor/cg1997.htm	C++	Software for multi-scale image analysis: the normalised, optimised anisotropic wavelet coefficient method. Darrozes (1997)
NOAWC pages http://renass.u-strasbg.fr/ ~philippe/Couleur/ coul_titre.html		

8.6 Chapter 6: Grain spatial distributions and relations

Big *R* calculations http://geologie.uqac.ca/ %7Emhiggins/big_r.zip	Win	Michael D. Higgins, UQAC

References

Agrawal, Y., McCave, I. N. & Riley, J. (1991). Laser diffraction size analysis. In J. P. M. Syvitski, ed., *Principles, Methods, and Application of Particle Size Analysis*. New York: Cambridge University Press.

Akesson, U., Lindqvist, J. E., Goransson, M. & Stigh, J. (2001). Relationship between texture and mechanical properties of granites, Central Sweden, by the use of image-analysing technique. *Bulletin of Engineering Geology and the Environment*, **60**, 277–84.

Akesson, U., Stigh, J., Lindqvist, J. E. & Goransson, M. (2003). The influence of foliation on the fragility of granitic rocks, image analysis and quantitative microscopy. *Engineering Geology*, **68**, 275–88.

Allen, S. R. & McPhie, J. (2003). Phenocryst fragments in rhyolitic lavas and lava domes. *Journal of Volcanology and Geothermal Research*, **126**, 263–83.

Anderson, A. T. (1983). Oscillatory zoning of plagioclase: Nomarski interference contrast microscopy of etched polished sections. *American Mineralogist*, **68**, 125–9.

Arbaret, L., Fernandez, A., Jezek, J., Ildefonse, B., Launeau, P. & Diot, H. (2000). Analogue and numerical modelling of shape fabrics: application to strain and flow determination in magmas. *Transactions of the Royal Society of Edinburgh-Earth Sciences*, **91**, 97–109.

Armienti, P. & Tarquini, S. (2002). Power law olivine crystal size distributions in lithospheric mantle xenoliths. *Lithos*, **65**, 273–85.

Armienti, P., Pareschi, M. T., Innocenti, F. & Pompilio, M. (1994). Effects of magma storage and ascent on the kinetics of crystal growth. The case of the 1991–93 Mt. Etna eruption. *Contributions to Mineralogy and Petrology*, **115**, 402–14.

ASTM (1992). ASTM E930–92e1 Standard test methods for estimating the largest grain observed in a metallographic section. Philadelphia, PA: American Society for Testing Materials.

ASTM (1996). ASTM E112–96 Standard test methods for determining average grain size. Philadelphia, PA: American Society for Testing Materials.

ASTM (1997). ASTM E1382–97 Standard test method for determining average grain size using semiautomatic and automatic image analysis. Philadelphia, PA: American Society for Testing Materials.

Baronnet, A. (1984). Growth kinetics of the silicates. A review of basic concepts. *Fortschritte der mineralogie*, **62**, 187–232.

Barrett, P. J. (1980). The shape of rock particles, a critical review. *Sedimentology*, **27**, 291–303.

Bateman, P. C. & Chappell, B. W. (1979). Crystallisation, fractionation, and solidification of the Tuolumne Intrusive series, Yosemite National Park, California. *Geological Society of America Bulletin*, **90**, 465–82.

Beere, W. (1975). Unifying theory of stability of penetrating liquid-phases and sintering pores. *Acta Metallurgica*, **23**, 131–8.

Ben Ismail, W. & Mainprice, D. (1998). An olivine fabric database: an overview of upper mantle fabrics and seismic anisotropy. *Tectonophysics*, **296**, 145–57.

Benn, K. & Mainprice, D. (1989). An interactive program for determination of plagioclase crystal axes orientations from U-stage measurements – an aid for petrofabric studies. *Computers & Geosciences*, **15**, 1127–42.

Bennema, P., Meekes, H. & Enckevort, V. (1999). Crystal growth and morphology: A multi-faceted approach. In B. Jamtveit & P. Meakin, eds., *Growth, Dissolution, and Pattern Formation in Geosystems*. Dordrecht: Boston, pp. 21–64.

Berger, A. (2004). An improved equation for crystal size distribution in second-phase influenced aggregates. *American Mineralogist*, **89**, 126–31.

Berger, A. & Herwegh, M. (2004). Grain coarsening in contact metamorphic carbonates: effects of second-phase particles, fluid flow and thermal perturbations. *Journal of Metamorphic Geology*, **22**, 459–74.

Berger, A. & Roselle, G. (2001). Crystallization processes in migmatites. *American Mineralogist*, **86**, 215–24.

Bevington, P. R. & Robinson, D. K. (2003). *Data Reduction and Error Analysis for the Physical Sciences*. Boston, MA: McGraw-Hill.

Bindeman, I. & Valley, J. W. (2001). Low-δ^{18}O rhyolites from Yellowstone: Magmatic evolution based on analyses of zircons and individual phenocrysts. *Journal of Petrology*, **42**, 1491–517.

Bindeman, I. N. (2003). Crystal sizes in evolving silicic magma chambers. *Geology*, **31**, 367–70.

Blanchard, J.-P., Boyer, P. & Gagny, C. (1979). Un nouveau critère de sens de mise en place dans une caisse filonienne: Le 'pincement' des minéraux aux epontes (Orientation des minéraux dans un magma en écoulement). *Tectonophysics*, **53**, 1–25.

Blower, J. D., Keating, J. P., Mader, H. M. & Phillips, J. C. (2001). Inferring volcanic degassing processes from vesicle size distributions. *Geophysical Research Letters*, **28**, 347–50.

Blower, J. D., Keating, J. P., Mader, H. M. & Phillips, J. C. (2002). The evolution of bubble size distributions in volcanic eruptions. *Journal of Volcanology and Geothermal Research*, **120**, 1–23.

Blumenfeld, P. & Bouchez, J. L. (1988). Shear criteria in granite and migmatite deformed in the magmatic and solid states. *Journal of Structural Geology*, **10**, 361–72.

Boorman, S., Boudreau, A. & Kruger, F. J. (2004). The lower zone-critical zone transition of the Bushveld complex: A quantitative textural study. *Journal of Petrology*, **45**, 1209–35.

Borradaile, G. J. (1988). Magnetic susceptibility, petrofabrics and strain. *Tectonophysics*, **156**, 1–20.

Borradaile, G. J. & Gauthier, D. (2003). Interpreting anomalous magnetic fabrics in ophiolite dikes. *Journal of Structural Geology*, **25**, 171–82.

Borradaile, G. J. & Henry, B. (1997). Tectonic applications of magnetic susceptibility and its anisotropy. *Earth-Science Reviews*, **42**, 49–93.

Bouchez, J. L. (1997). Granite is never isotropic: an introduction to AMS studies of granitic rocks. In J. L. Bouchez, D. H. W. Hutton & W. E. Stephens, eds., *Granite: From Segregation of Melt to Emplacement Fabrics*. Dordrecht: Kluwer Academic Publishers.

Bouchez, J. L. (2000). Magnetic susceptibility anisotropy and fabrics in granites. *Comptes rendus de l'Academie des Sciences*, **330**, 1–14.

Boudreau, A. E. (1987). Pattern forming during crystallisation and the formation of fine-scale layering. In I. Parsons, ed., *Origins of Igneous Layering*. Dordrecht: D. Reidel, pp. 453–71.

Boudreau, A. E. (1995). Crystal aging and the formation of fine-scale igneous layering. *Mineralogy and Petrology*, **54**, 55–69.

Bowman, E. T., Soga, K. & Drummond, W. (2001). Particle shape characterisation using Fourier descriptor analysis. *Geotechnique*, **51**, 545–54.

Bozhilov, K. N., Green, H. W. & Dobrzhinetskaya, L. F. (2003). Quantitative 3D measurement of ilmenite abundance in Alpe Arami olivine by confocal microscopy: Confirmation of high-pressure origin. *American Mineralogist*, **88**, 596–603.

Brandeis, G. & Jaupart, C. (1987). The kinetics of nucleation and crystal growth and scaling laws for magmatic crystallisation. *Contributions to Mineralogy and Petrology*, **96**, 24–34.

Brandon, D. G. & Kaplan, W. D. (1999). *Microstructural Characterization of Materials*. Chichester, NY: J. Wiley and Sons.

Brantley, S. L. & Mellott, N. P. (2000). Surface area and porosity of primary silicate minerals. *American Mineralogist*, **85**, 1767–83.

Brantley, S. L., White, A. F. & Hodson, M. E. (1999). Surface area of primary silicate minerals. In B. Jamtveit & P. Meakin, eds., *Growth, Dissolution, and Pattern Formation in Geosystems*. Dordrecht: Boston, pp. 291–326.

Bryon, D. N., Atherton, M. P. & Hunter, R. H. (1995). The interpretation of granitic textures from serial thin sectioning, image-analysis and 3-dimensional reconstruction. *Mineralogical Magazine*, **59**, 203–11.

Bulau, J. R., Waff, H. S. & Tyburczy, J. A. (1979). Mechanical and thermodynamic constraints on fluid distribution in partial melts. *Journal of Geophysical Research*, **84**, 6102–8.

Bunge, H. J. (1982). *Texture Analysis in Materials Science*. London, UK: Butterworths.

Burnard, P. (1999). Eruption dynamics of 'popping rock' from vesicle morphologies. *Journal of Volcanology and Geothermal Research*, **92**, 247–58.

Cabane, H., Laporte, D. & Provost, A. (2001). Experimental investigation of the kinetics of Ostwald ripening of quartz in silicic melts. *Contributions to Mineralogy and Petrology*, **142**, 361–73.

Cabri, L. & Vaughan, D., eds., (1998). *Modern Approaches to Ore and Environmental Mineralogy*. Short Course Series, 27. Ottawa: Mineralogical Association of Canada.

Canon-Tapia, E. & Castro, J. (2004). AMS measurements on obsidian from the Inyo Domes, CA: a comparison of magnetic and mineral preferred orientation fabrics. *Journal of Volcanology and Geothermal Research*, **134**, 169–82.

Canon-Tapia, E., Walker, G. P. L. & Herrero-Bervera, E. (1997). The internal structure of lava flows – Insights from AMS measurements II: Hawaiian pahoehoe, toothpaste lava and a'a. *Journal of Volcanology and Geothermal Research*, **76**, 19–46.

Capaccioni, B. & Sarocchi, D. (1996). Computer-assisted image analysis on clast shape fabric from the Orvieto-Bagnoregio ignimbrite (Vulsini District, central Italy): Implications on the emplacement mechanisms. *Journal of Volcanology and Geothermal Research*, **70**, 75–90.

Capaccioni, B., Valentini, L., Rocchi, M. B. L., Nappi, G. & Sarocchi, D. (1997). Image analysis and circular statistics for shape-fabric analysis: Applications to lithified ignimbrites. *Bulletin of Volcanology*, **58**, 501–14.

Carey, S., Maria, A. & Sigurdsson, H. (2000). Use of fractal analysis for discrimination of particles from primary and reworked jokulhlaup deposits in SE Iceland. *Journal of Volcanology and Geothermal Research*, **104**, 65–80.

Carlson, W. D. (1999). The case against Ostwald ripening of porphyroblasts. *Canadian Mineralogist*, **37**, 403–13.

Carlson, W. D., Denison, C. & Ketcham, R. A. (1995). Controls on the nucleation and growth of porphyroblasts: Kinetics from natural textures and numerical models. *Geological Journal*, **30**, 207–25.

Cashman, K. (1986). Crystal size distributions in igneous and metamorphic rocks. Baltimore, MA: Johns Hopkins University.

Cashman, K. & Blundy, J. (2000). Degassing and crystallization of ascending andesite and dacite. *Philosophical Transactions of the Royal Society of London Series A (Mathematical Physical and Engineering Sciences)*, **358**, 1487–513.

Cashman, K. V. (1988). Crystallisation of Mount St. Helens dacite; a quantitative textural approach. *Bulletin of Volcanology*, **50**, 194–209.

Cashman, K. V. (1990). Textural constraints on the kinetics of crystallization of igneous rocks. In J. Nicholls & J. K. Russell, eds., *Modern Methods of Igneous Petrology: Understanding Magmatic Processes*. Washington DC: Mineralogical Society of America, pp. 259–314.

Cashman, K. V. (1992). Groundmass crystallisation of Mount St Helens dacite 1980–1986: a tool for interpreting shallow magmatic processes. *Contributions to Mineralogy and Petrology*, **109**, 431–49.

Cashman, K. V. (1993). Relationship between plagioclase crystallisation and cooling rate in basaltic melts. *Contributions to Mineralogy and Petrology*, **113**, 126–42.

Cashman, K. V. & Ferry, J. M. (1988). Crystal size distribution (CSD) in rocks and the kinetics and dynamics of crystallization III. Metamorphic crystallization. *Contributions to Mineralogy and Petrology*, **99**, 410–15.

Cashman, K. V. & Marsh, B. D. (1988). Crystal size distribution (CSD) in rocks and the kinetics and dynamics of crystallisation II. Makaopuhi lava lake. *Contributions to Mineralogy and Petrology*, **99**, 292–305.

Cashman, K. V., Thornber, C. & Kauahikaua, J. P. (1999). Cooling and crystallization of lava in open channels, and the transition of pahoehoe lava to a'a. *Bulletin of Volcanology*, **61**, 306–23.

Castro, J. M., Cashman, K. V. & Manga, M. (2003). A technique for measuring 3D crystal-size distributions of prismatic microlites in obsidian. *American Mineralogist*, **88**, 1230–40.

Chacron, M. & L'Heureux, I. (1999). A new model of periodic precipitation incorporating nucleation, growth and ripening. *Physics Letters A*, **263**, 70–7.

Chayes, F. (1950). On the bias of grain-size measurements made in thin section. *Journal of Geology*, **58**, 156–60.

Cheadle, M. J., Elliott, M. T. & McKenzie, D. (2004). Percolation threshold and permeability of crystallizing igneous rocks: The importance of textural equilibrium. *Geology*, **32**, 757–60.

Christiansen, C. & Hartman, D. (1991). Principles, methods, and application of particle size analysis. In J. P. M. Syvitski, ed., *Principles, Methods, and Application of Particle Size Analysis*. New York: Cambridge University Press, pp. 237–48.

Clark, A. H., Pearce, T. H., Roeder, P. L. & Wolfson, I. (1986). Oscillatory zoning and other microstructures in magmatic olivine and augite; Nomarski interference contrast observations on etched polished surfaces. *American Mineralogist*, **71** 734–41.

Cloetens, P., Ludwig, W., Boller, E., Peyrin, F., Schlenker, M. & Baruchel, J. (2002). 3D imaging using coherent synchrotron radiation. *Image Analysis and Stereology*, **21** (suppl. 1), S75–S86.

Cmiral, M., FitzGerald, J. D., Faul, U. H. & Green, D. H. (1998). A close look at dihedral angles and melt geometry in olivine-basalt aggregates: a TEM study. *Contributions to Mineralogy and Petrology*, **130**, 336–45.

Coakley, J. & Syvitski, J. P. M. (1991). Sedigraph technique. In J. P. M. Syvitski, ed., *Principles, Methods, and Application of Particle Size Analysis*. New York: Cambridge University Press, pp. 129–42.

Coogan, L. A., Thompson, G. & MacLeod, C. J. (2002). A textural and geochemical investigation of high level gabbros from the Oman ophiolite: implications for the role of the axial magma chamber at fast-spreading ridges. *Lithos*, **63**, 67–82.

Costa, L. F. & Cesar, R. M. (2001). *Shape Analysis and Classification: Theory and Practice*. Boca Raton, FL: CRC Press.

Craig, D. B. (1961). The Benford Plate. *American Mineralogist*, **46**, 757–8.

Cressie, N. (1991). *Statistics for Spatial Data*. New York: Wiley-Interscience.

Cruden, A. R. & Launeau, P. (1994). Structure, magnetic fabric and emplacement of the Archean Lebel stock, SW Abitibi greenstone belt. *Journal of Structural Geology*, **16**, 677–91.

Dana, J. D., Dana, E. S., Palache, C., Berman, H. M. & Frondel, C. (1944). *The System of Mineralogy of James Dwight Dana and Edward Salisbury Dana, Yale University, 1837–1892*. New York: J. Wiley and Sons, London: Chapman and Hall.

Daniel, C. G. & Spear, F. S. (1999). The clustered nucleation and growth processes of garnet in regional metamorphic rocks from north-west Connecticut, USA. *Journal of Metamorphic Geology*, **17**, 503–20.

Darrozes, J., Gaillot, P., De Saint-Blanquat, M. & Bouchez, J. L. (1997). Software for multi-scale image analysis: The normalized optimized anisotropic wavelet coefficient method. *Computers & Geosciences*, **23**, 889–95.

Deakin, A. S. & Boxer, G. L. (1989). Argyle AK1 diamond size distribution; the use of fine diamonds to predict the occurrence of commercial size diamonds. In J. Ross, A. L. Jaques, J. Ferguson, D. H. Green, S. Y. O'Reilly, R. V. Danchin & A. J. A. Janse, eds., *Fourth International Kimberlite Conference*. Sydney: Geological Society of Australia, pp. 1117–22.

DeHoff, R. T. (1984). Generalized microstructural evolution by interface controlled coarsening. *Acta Metallurgica*, **32**, 43–7.

DeHoff, R. T. (1991). A geometrically general theory of diffusion controlled coarsening. *Acta Metallurgica et Materialia*, **39**, 2349–60.

Delesse, M. A. (1847). Procedé mécanique pour déterminer la composition des roches. *Comptes rendus de l'Académie des Sciences (Paris)*, **25**, 544–5.

Dellino, P. & Liotino, G. (2002). The fractal and multifractal dimension of volcanic ash particles contour: a test study on the utility and volcanological relevance. *Journal of Volcanology and Geothermal Research*, **113**, 1–18.

Dillon, C. G., Carey, P. F. & Worden, R. H. (2001). Fractscript: A macro for calculating the fractal dimension of object perimeters in images of multiple objects. *Computers & Geosciences*, **27**, 787–94.

Diot, H., Bolle, O., Lambert, J. M., Launeau, P. & Duchesne, J. C. (2003). The Tellnes ilmenite deposit (Rogaland, South Norway): magnetic and petrofabric evidence for emplacement of a Ti-enriched noritic crystal mush in a fracture zone. *Journal of Structural Geology*, **25**, 481–501.

Donaldson, C. H. (1976). An experimental investigation of olivine morphology. *Contributions to Mineralogy and Petrology*, **57**, 187–213.

Dowty, E. (1980). Crystal growth and nucleation theory and the numerical simulation of igneous crystallization. In R. B. Hargraves, ed., *Physics of Magmatic Processes*. Princeton: Princeton University Press.

Drolon, H., Hoyez, B., Druaux, F. & Faure, A. (2003). Multiscale roughness analysis of particles: Application to the classification of detrital sediments. *Mathematical Geology*, **35**, 805–17.

Druitt, T. H., Edwards, L., Mellors, R. M., *et al.* (1999). *Santorini Volcano*. London, UK: Geological Society. Memoir 19

Dunbar, N. W., Cashman, K. V. & Dupre, R. (1994). Crystallization processes of anorthoclase phenocrysts in the Mount Erebus magmatic system; evidence from crystal composition, crystal size distributions, and volatile contents of melt inclusions. In P. R. Kyle, ed., *Volcanological and Environmental Studies of Mount Erebus, Antarctica*. Washington DC: American Geophysical Union, pp. 129–46.

Duyster, J. & Stockhert, B. (2001). Grain boundary energies in olivine derived from natural microstructures. *Contributions to Mineralogy and Petrology*, **140**, 567–76.

Eberl, D. D., Drits, V. A. & Srodon, J. (1998). Deducing growth mechanisms for minerals from the shapes of crystal size distributions. *American Journal of Science*, **298**, 499–533.

Eberl, D. D., Kile, D. E. & Drits, V. A. (2002). On geological interpretations of crystal size distributions: Constant vs. proportionate growth. *American Mineralogist*, **87**, 1235–41.

Ehrlich, R. & Weinberg, B. (1970). An exact method for characterization of grain shape. *Journal of Sedimentary Petrology*, **40**, 205–12.

Eisenhour, D. D. (1996). Determining chondrule size distributions from thin-section measurements. *Meteoritics & Planetary Science*, **31**, 243–8.

Elliott, M. T. & Cheadle, M. J. (1997). On the identification of textural disequilibrium in rocks using dihedral angle measurements: Reply. *Geology*, **25**, 1055.

Elliott, M. T., Cheadle, M. J. & Jerram, D. A. (1997). On the identification of textural equilibrium in rocks using dihedral angle measurements. *Geology*, **25**, 355–8.

Emmons, R. C. (1964). *The Universal Stage (With Five Axes of Rotation)*. New York: Geological Society of America.

Epstein, B. (1947). The mathematical description of certain breakage mechanisms leading to the logarithmico-normal distribution. *Journal of the Franklin Institute*, **244**, 471–7.

Ernst, R. E. & Baragar, W. R. A. (1992). Evidence from magnetic fabric for the flow pattern of magma in the Mackenzie giant radiating dyke swarm. *Nature*, **356**, 511–13.

Ersoy, A. & Waller, M. D. (1995). Textural characterization of rocks. *Engineering Geology*, **39**, 123–36.

Exner, H. (2004). Stereology and 3D microscopy: Useful alternatives or competitors in the quantitative analysis of microstructures? *Image Analysis and Stereology*, **23**, 73–82.

Faul, U. (1997). The permeability of partially molten upper mantle rocks from experiments and percolation theory. *Journal of Geophysical Research*, **102**, 10299–311.

Faure, F., Trolliard, G., Nicollet, C. & Montel, J. M. (2003). A developmental model of olivine morphology as a function of the cooling rate and the degree of undercooling. *Contributions to Mineralogy and Petrology*, **145**, 251–63.

Flinn, D. (1962). On folding during three-dimensional progressive deformation. *Geological Society of London Quarterly Journal*, **118**, 385–433.

Friedman, G. M. (1958). Determination of sieve-size distributions from thin-section data for sedimentary petrological studies. *Journal of Geology*, **66**, 394–416.

Fueten, F. (1997). A computer-controlled rotating polarizer stage for the petrographic microscope. *Computers & Geosciences*, **23**, 203–8.

Fueten, F. & Goodchild, J. S. (2001). Quartz c-axes orientation determination using the rotating polarizer microscope. *Journal of Structural Geology*, **23**, 895–902.

Gaillot, P., Darrozes, J., de Saint Blanquat, M. & Ouillon, G. (1997). The normalized optimized anisotropic wavelet coefficient (NOAWC) method; an image processing tool for multiscale analysis of rock fabric. *Geophysical Research Letters*, **24**, 1819–22.

Gaillot, P., Darrozes, J. & Bouchez, J. L. (1999). Wavelet transform: a future of rock fabric analysis? *Journal of Structural Geology*, **21**, 1615–21.

Galwey, A. K. & Jones, K. A. (1963). An attempt to determine the mechanism of a natural mineral-forming reaction from examination of the products. *Journal of the Chemical Society (London)*, 5681–6.

Galwey, A. K. & Jones, K. A. (1966). Crystal size frequency distribution of garnets in some analysed metamorphic rocks from Mallaig, Inverness, Scotland. *Geological Magazine*, **103**, 143–52.

Gaonac'h, H., Lovejoy, S., Stix, J. & Scherzter, D. (1996a). A scaling growth model for bubbles in basaltic lava flows. *Earth and Planetary Science Letters*, **139**, 395–409.

Gaonac'h, H., Stix, J. & Lovejoy, S. (1996b). Scaling effects on vesicle shape, size and heterogeneity of lavas from Mount Etna. *Journal of Volcanology and Geothermal Research*, **74**, 131–52.

Gaonac'h, H., Lovejoy, S. & Schertzer, D. (2003). Percolating magmas and explosive volcanism. *Geophysical Research Letters*, **30**, 1559.

Gardner, J. E. & Denis, M.-H. (2004). Heterogeneous bubble nucleation on Fe-Ti oxide crystals in high-silica rhyolitic melts. *Geochimica et Cosmochimica Acta*, **68**, 3587–97.

Gardner, J. E., Hilton, M. & Carroll, M. R. (1999). Experimental constraints on degassing of magma: isothermal bubble growth during continuous decompression from high pressure. *Earth and Planetary Science Letters*, **168**, 201–18.

Garrido, C. J., Kelemen, P. B. & Hirth, G. (2001). Variation of cooling rate with depth in lower crust formed at an oceanic spreading ridge; plagioclase crystal size distributions in gabbros from the Oman Ophiolite. *Geochemistry Geophysics Geosystems*, doi: 10.1029/2000GC000136.

Gee, J. S., Meurer, W. P., Selkin, P. A. & Cheadle, M. J. (2004). Quantifying three-dimensional silicate fabrics in cumulates using cumulative distribution functions. *Journal of Petrology*, **45**, 1983–2009.

Geoffroy, L., Callot, J. P., Aubourg, C. & Moreira, M. (2002). Magnetic and plagioclase linear fabric discrepancy in dykes: a new way to define the flow vector using magnetic foliation. *Terra Nova*, **14**, 183–90.

Ghiorso, M. S. & Sack, R. O. (1995). Chemical mass-transfer in magmatic processes. 4. A revised and internally consistent thermodynamic model for the interpolation and extrapolation of liquid-solid equilibria in magmatic systems at elevated-temperatures and pressures. *Contributions to Mineralogy and Petrology*, **119**, 197–212.

Gingras, M. K., MacMillan, B. & Balcom, B. J. (2002). Visualizing the internal physical characteristics of carbonate sediments with magnetic resonance imaging and petrography. *Bulletin of Canadian Petroleum Geology*, **50**, 363–9.

Goodchild, J. S. & Fueten, F. (1998). Edge detection in petrographic images using the rotating polarizer stage. *Computers & Geosciences*, **24**, 745–51.

Goodrich, C. A. (2003). Petrogenesis of olivine-phyric shergottites Sayh Al Uhaymir 005 and elephant moraine A79001 lithology A. *Geochimica et Cosmochimica Acta*, **67**, 3735–72.

Gray, N. H. (1970). Crystal growth and nucleation in two large diabase dykes. *Canadian Journal of Earth Sciences*, **7**, 366–75.

Gray, N. H., Philpotts, A. R. & Dickson, L. D. (2003). Quantitative measures of textural anisotropy resulting from magmatic compaction illustrated by a sample from the Palisades sill, New Jersey. *Journal of Volcanology and Geothermal Research*, **121**, 293–312.

Gregg, S. J. & Sing, K. S. W. (1982). *Adsorption, Surface Area, and Porosity*. London: Academic Press.

Gregoire, V., Darrozes, J., Gaillot, P., Nedelec, A. & Launeau, P. (1998). Magnetite grain shape fabric and distribution anisotropy vs rock magnetic fabric: a three-dimensional case study. *Journal of Structural Geology*, **20**, 937–44.

Greshake, A., Fritz, J. & Stoffler, D. (2004). Petrology and shock metamorphism of the olivine-phyric shergottite Yamato 980459: Evidence for a two-stage cooling and a single-stage ejection history. *Geochimica et Cosmochimica Acta*, **68**, 2359–77.

Gualda, G., Cook, D., Chopra, R. *et al.* (2004). Fragmentation, nucleation and migration of crystals and bubbles in the Bishop Tuff rhyolitic magma. *Transactions of the Royal Society of Edinburgh-Earth Sciences*, **95**, 375–90.

Habesch, S. M. (2000). Electron backscattered diffraction analyses combined with environmental scanning electron microscopy: potential applications for non-conducting, uncoated mineralogical samples. *Materials Science and Technology*, **16**, 1393–8.

Hammer, J. E., Cashman, K. V., Hoblitt, R. P. & Newman, S. (1999). Degassing and microlite crystallization during pre-climactic events of the 1991 eruption of Mt. Pinatubo, Philippines. *Bulletin of Volcanology*, **60**, 355–80.

Hanchar, J. M. & Hoskin, P. W. O., eds., (2003). *Zircon*. Reviews in Mineralogy and Geochemistry, 53. Washington, DC: Mineralogical Society of America.

Harvey, P. K. & Laxton, R. R. (1980). The estimate of finite strain from the orientation distribution of passively deformed linear markers: eigenvalue relationships. *Tectonophysics*, **70**, 285–307.

Heilbronner, P. R. & Bruhn, D. (1998). The influence of three-dimensional grain size distributions on the rheology of polyphase rocks. *Journal of Structural Geology*, **20**, 695–705.

Heilbronner, R. (2002). Analysis of bulk fabrics and microstructure variations using tesselations of autocorrelation functions. *Computers & Geosciences*, **28**, 447–55.

Heilbronner, R. P. (1992). The autocorrelation function – an image-processing tool for fabric analysis. *Tectonophysics*, **212**, 351–70.

Heilbronner, R. P. & Pauli, C. (1993). Integrated spatial and orientation analysis of quartz c-axes by computer-aided microscopy. *Journal of Structural Geology*, **15**, 369–82.

Herring, C. (1951a). Some theorems on the free energies of crystal surfaces. *Physics Reviews*, **82**, 87–93.

Herring, C. (1951b). Surface tension as a motivation for sintering. In W. Kingston, ed., *Physics of Powder Metallurgy*. New York: McGraw-Hill.

Herwegh, M. (2000). A new technique to automatically quantify microstructures of fine grained carbonate mylonites: two-step etching combined with SEM imaging and image analysis. *Journal of Structural Geology*, **22**, 391–400.

Herwegh, M., de Bresser, J. & ter Heege, J. (2005). Combining natural microstructures with composite flow laws: an improved approach for the extrapolation of lab data to nature. *Journal of Structural Geology*, **27**, 503–21.

Hext, G. (1963). The estimation of second-order tensors, with related tests and designs. *Biometrika*, **50**, 353–73.

Heyraud, J. C. & Metois, J. J. (1987). Equilibrium shape of an ionic-crystal in equilibrium with its vapor (NaCl). *Journal of Crystal Growth*, **84**, 503–8.

Higgins, M. D. (1991). The origin of laminated and massive anorthosite, Sept Iles intrusion, Quebec, Canada. *Contributions to Mineralogy and Petrology*, **106**, 340–54.

Higgins, M. D. (1994). Determination of crystal morphology and size from bulk measurements on thin sections: numerical modelling. *American Mineralogist*, **79**, 113–19.

Higgins, M. D. (1996a). Crystal size distributions and other quantitative textural measurements in lavas and tuff from Mt Taranaki (Egmont volcano), New Zealand. *Bulletin of Volcanology*, **58**, 194–204.

Higgins, M. D. (1996b). Magma dynamics beneath Kameni volcano, Greece, as revealed by crystal size and shape measurements. *Journal of Volcanology and Geothermal Research*, **70**, 37–48.

Higgins, M. D. (1998). Origin of anorthosite by textural coarsening: Quantitative measurements of a natural sequence of textural development. *Journal of Petrology*, **39**, 1307–25.

Higgins, M. D. (1999). Origin of megacrysts in granitoids by textural coarsening: a crystal size distribution (CSD) study of microcline in the Cathedral Peak granodiorite, Sierra Nevada, California. In C. Fernandez & A. Castro, eds., *Understanding Granites: Integrating Modern and Classical Techniques. Special Publication 158*. London: Geological Society of London, pp. 207–19.

Higgins, M. D. (2000). Measurement of crystal size distributions. *American Mineralogist*, **85**, 1105–16.

Higgins, M. D. (2002a). Closure in crystal size distributions (CSD), verification of CSD calculations, and the significance of CSD fans. *American Mineralogist*, **87**, 171–5.

Higgins, M. D. (2002b). A crystal size-distribution study of the Kiglapait layered mafic intrusion, Labrador, Canada: evidence for textural coarsening. *Contributions to Mineralogy and Petrology*, **144**, 314–30.

Higgins, M. D. (2005). A new model for the structure of the Sept Iles intrusive suite, Canada. *Lithos*, **83**, 199–213.

Higgins, M. D. (2006). Use of appropriate diagrams to determine if crystal size distributions (CSD) are dominantly semi-logarithmic, lognormal or fractal (scale invariant). *Journal of Volcanology and Geothermal Research*.

Higgins, M. D. & Roberge, J. (2003). Crystal size distribution (CSD) of plagioclase and amphibole from Soufriere Hills volcano, Montserrat: Evidence for dynamic crystallisation / textural coarsening cycles. *Journal of Petrology*, **44**, 1401–11.

Hiraga, T., Nishikawa, O., Nagase, T. & Akizuki, M. (2001). Morphology of intergranular pores and wetting angles in pelitic schists studied by transmission electron microscopy. *Contributions to Mineralogy and Petrology*, **141**, 613–22.

Hiraga, T., Nishikawa, O., Nagase, T., Akizuki, M. & Kohlstedt, D. L. (2002). Interfacial energies for quartz and albite in pelitic schist. *Contributions to Mineralogy and Petrology*, **143**, 664–72.

Holness, M. B. (1993). Temperature and pressure-dependence of quartz aqueous fluid dihedral angles – the control of adsorbed H_2O on the permeability of quartzites. *Earth and Planetary Science Letters*, **117**, 363–77.

Holness, M. B. (2005). Spatial constraints on magma chamber replenishment events from textural observations of cumulates: the Rum layered intrusion, Scotland. *Journal of Petrology*, **46**, 1585–601.

Holness, M. B. & Siklos, S. T. C. (2000). The rates and extent of textural equilibration in high-temperature fluid-bearing systems. *Chemical Geology*, **162**, 137–53.

Holness, M. B., Cheadle, M. C. & McKenzie, D. (2005). On the use of changes in dihedral angle to decode late-stage textural evolution in cumulates. *Journal of Petrology*, **46**, 1565–83.

Houghton, B. F. & Wilson, C. J. N. (1989). A vesicularity index for pyroclastic deposits. *Bulletin of Volcanology*, **51**, 451–62.

Howard, V. & Reed, M. G. (1998). *Unbiased Stereology: Three-Dimensional Measurement in Microscopy*. Oxford, UK: Bios Scientific Publishers, New York: Springer.

Howarth, D. & Rowlands, J. (1986). Development of an index to quantify rock textures for quantitative assessment of intact rock properties. *Geotechnical Testing Journal*, **9**, 169–79.

Hunter, R. H. (1987). Textural Equilibrium in Layered Igneous Rocks. In I. Parsons, ed., *Origins of Igneous Layering*. Dordrecht: D. Reidel, pp. 473–503.

Hunter, R. H. (1996). Textural Development in Cumulate Rocks. In R. G. Cawthorn, ed., *Layered Intrusions*. Amsterdam: Elsevier, pp. 77–101.

Hurwitz, S. & Navon, O. (1994). Bubble nucleation in rhyolitic melts – experiments at high-pressure, temperature, and water-content. *Earth and Planetary Science Letters*, **122**, 267–80.

Hutchison, C. S. (1974). *Laboratory Handbook of Petrographic Techniques*. Hoboken NY: John Wiley & Sons.

Iezzi, G. & Ventura, G. (2002). Crystal fabric evolution in lava flows: results from numerical simulations. *Earth and Planetary Science Letters*, **200**, 33–46.

Ihinger, P. D. & Zink, S. I. (2000). Determination of relative growth rates of natural quartz crystals. *Nature*, **404**, 865–9.

Ikeda, S., Toriumi, M., Yoshida, H. & Shimizu, I. (2002). Experimental study of the textural development of igneous rocks in the late stage of crystallization: the importance of interfacial energies under non-equilibrium conditions. *Contributions to Mineralogy and Petrology*, **142**, 397–415.

Ikeda, S., Nakano, T., Tsuchiyama, A., Uesugi, K., Suzuki, Y., Nakamura, K., Nakashima, Y. & Yoshida, H. (2004). Nondestructive three-dimensional element-concentration mapping of a Cs-doped partially molten granite by X-ray computed tomography using synchrotron radiation. *American Mineralogist*, **89**, 1304–13.

Ildefonse, B., Launeau, P. & Bouchez, J.-L. (1992). Effect of mechanical interactions on the development of shape preferred orientations: a two-dimensional experimental approach. *Journal of Structural Geology*, **14**, 73–83.

Jackson, E. D. (1961). Primary textures and mineral associations in the ultramafic zone of the Stillwater complex, Montana. *United States Geological Survey Professional Paper*, **358**.

Jackson, M. (1991). Anisotropy of magnetic remanence – a brief review of mineralogical sources, physical origins, and geological applications, and comparison with susceptibility anisotropy. *Pure and Applied Geophysics*, **136**, 1–28.

Jackson, M., Gruber, W., Marvin, J. & Banerjee, S. K. (1988). Partial anhysteretic remanence and its anisotropy – applications and grainsize-dependence. *Geophysical Research Letters*, **15**, 440–3.

Jamtveit, B. & Meakin, P., eds., (1999). *Growth, Dissolution, and Pattern Formation in Geosystems*. Dordrecht: Kluwer Academic Publishers.

Jelinek, V. (1978). Statistical processing of anisotropy of magnetic-susceptibility measured on groups of specimens. *Studia Geophysica et Geodaetica*, **22**, 50–62.

Jerram, D. A. (2001). Visual comparators for degree of grain-size sorting in two and three-dimensions. *Computers & Geoscience*, **27**, 485–92.

Jerram, D. A. & Cheadle, M. J. (2000). On the cluster analysis of grains and crystals in rocks. *American Mineralogist*, **85**, 47–67.

Jerram, D. A., Cheadle, M. J., Hunter, R. H. & Elliott, M. T. (1996). The spatial distribution of grains and crystals in rocks. *Contributions to Mineralogy and Petrology*, **125**, 60–74.

Jerram, D. A., Cheadle, M. J. & Philpotts, A. R. (2003). Quantifying the building blocks of igneous rocks: Are clustered crystal frameworks the foundation? *Journal of Petrology*, **44**, 2033–51.

Ji, S. C., Zhao, X. O. & Zhao, P. L. (1994). On the measurement of plagioclase lattice preferred orientations. *Journal of Structural Geology*, **16**, 1711–18.

Jillavenkateas, A., Dapkunas, S. J. & Lum, L.-S. H. (2001). *Particle Size Characterisation: NIST Recommended Practice Guide*. Washington DC, USA: National Institute of Standards and Technology.

Johnson, M. R. (1994). Thin-section grain-size analysis revisited. *Sedimentology*, **41**, 985–99.

Jung, H. & Karato, S. (2001). Effects of water on dynamically recrystallized grain-size of olivine. *Journal of Structural Geology*, **23**, 1337–44.

Jurewicz, S. R. & Jurewicz, A. J. G. (1986). Distribution of apparent angles on random sections with emphasis on dihedral angle measurements. *Journal of Geophysical Research-Solid Earth and Planets*, **91**, 9277–82.

Jurewicz, S. R. & Watson, E. B. (1985). The distribution of partial melt in a granitic system – the application of liquid-phase sintering theory. *Geochimica et Cosmochimica Acta*, **49**, 1109–21.

Karato, S. & Wenk, H.-R. (2002). *Plastic Deformation of Minerals and Rocks*. Washington, DC: Mineralogical Society of America.

Ketcham, R. A. (2005). Computational methods for quantitative analysis of three-dimensional features in geological specimens. *Geosphere*, **1**, 32–41.

Ketcham, R. A. & Carlson, W. D. (2001). Acquisition, optimization and interpretation of X-ray computed tomographic imagery; applications to the geosciences. *Computers & Geoscience*, **27**, 381–400.

Ketcham, R. A. & Ryan, T. M. (2004). Quantification and visualization of anisotropy in trabecular bone. *Journal of Microscopy-Oxford*, **213**, 158–71.

Kile, D. E., Eberl, D. D., Hoch, A. R. & Reddy, M. M. (2000). An assessment of calcite crystal growth mechanisms based on crystal size distributions. *Geochimica et Cosmochimica Acta*, **64**, 2937–50.

Klug, C. & Cashman, K. (1994). Vesiculation of May 18, 1980 Mount St. Helens magma. *Geology*, **22**, 468–72.

Klug, C., Cashman, K. V. & Bacon, C. R. (2002). Structure and physical characteristics of pumice from the climactic eruption of Mount Mazama (Crater Lake), Oregon. *Bulletin of Volcanology*, **64**, 486–501.

Knight, M. D., Walker, G. P. L., Ellwood, B. B. & Diehl, J. F. (1986). Stratigraphy, paleomagnetism, and magnetic fabric of the Toba Tuffs – constraints on the sources and eruptive styles. *Journal of Geophysical Research-Solid Earth and Planets*, **91**, 355–82.

Kocks, U. F., Wenk, H.-R. & Tomé, C. N. (2000). *Texture and Anisotropy: Preferred Orientations in Polycrystals and Their Effect on Material Properties*. Cambridge, UK: Cambridge University Press.

Kolmogorov, A. (1941). The lognormal law of distribution of particle sizes during crushing. *Doklady Akademii Nauk SSSR*, **31**, 99–101.

Kong, M. Y., Bhattacharya, R. N., James, C. & Basu, A. (2005). A statistical approach to estimate the 3D size distribution of spheres from 2D size distributions. *Geological Society of America Bulletin*, **117**, 244–9.

Kostov, I. & Kostov, R. I. (1999). *Crystal Habits of Minerals*. Sofia: Prof. Marin Drinov Academic Publishing House; Pensoft Publishers.

Kotov, S. & Berendsen, P. (2002). Statistical characteristics of xenoliths in the Antioch kimberlite pipe, Marshall county, northeastern Kansas. *Natural Resources Research*, **11**, 289–97.

Kouchi, A., Tsuchiyama, A. & Sunagawa, I. (1986). Effects of stirring on crystallization of basalt: Texture and element partitioning. *Contributions to Mineralogy and Petrology*, **93**, 429–38.

Kretz, R. (1966a). Grain-size distribution for certain metamorphic minerals in relation to nucleation and growth. *Journal of Geology*, **74**, 147–73.

Kretz, R. (1966b). Interpretation of shape of mineral grains in metamorphic rocks. *Journal of Petrology*, **7**, 68–94.

Kretz, R. (1969). On the spatial distribution of crystals in rocks. *Lithos*, **2**, 39–65.

Kretz, R. (1993). A garnet population in Yellowknife schist, Canada. *Journal of Metamorphic Geology*, **11**, 101–20.

Krug, H. J., Brandtstadter, H. & Jacob, K. H. (1996). Morphological instabilities in pattern formation by precipitation and crystallization processes. *Geologische Rundschau*, **85**, 19–28.

Kruhl, J. H. & Nega, M. (1996). The fractal shape of sutured quartz grain boundaries: Application as a geothermometer. *Geologische Rundschau*, **85**, 38–43.

Kruhl, J. H. & Peternell, M. (2002). The equilibration of high-angle grain boundaries in dynamically recrystallized quartz: the effect of crystallography and temperature. *Journal of Structural Geology*, **24**, 1125–37.

Lane, A. C. (1898). *Geological Report on Isle Royale, Michigan*. Lansing, MI: Michigan Geological Survey.

Laporte, D. & Provost, A. (2000). Equilibrium geometry of a fluid phase in a polycrystalline aggregate with anisotropic surface energies: Dry grain boundaries. *Journal of Geophysical Research-Solid Earth*, **105**, 25937–53.

Laporte, D. & Watson, E. B. (1995). Experimental and theoretical constraints on melt distribution in crustal sources – the effect of crystalline anisotropy on melt interconnectivity. *Chemical Geology*, **124**, 161–84.

Larsen, J. F. & Gardner, J. E. (2000). Experimental constraints on bubble interactions in rhyolite melts; implications for vesicle size distributions. *Earth and Planetary Science Letters*, **180**, 201–14.

Larsen, J. F., Denis, M. H. & Gardner, J. E. (2004). Experimental study of bubble coalescence in rhyolitic and phonolitic melts. *Geochimica et Cosmochimica Acta*, **68**, 333–44.

Larsen, L. & Poldervaart, A. (1957). Measurements and distribution of zircons in some granitic rocks of magmatic origin. *Mineralogical Magazine*, **31**, 544–64.

Lasaga, A. C. (1998). *Kinetic Theory in the Earth Sciences*. Princeton NJ: Princeton University Press.

Launeau, P. (2004). Mise en evidence des écoulments magmatiques par analyse d'images 2-D des distibutions 3-D d'orientations préférentielles de formes. *Bulletin de la Société Géologique de France*, **175**, 331–50.

Launeau, P. & Cruden, A. R. (1998). Magmatic fabric acquisition mechanisms in a syenite: results of a combined AMS and image analysis study. *Journal of Geophysical Research*, **103**, 5067–89.

Launeau, P. & Robin, P. Y. F. (1996). Fabric analysis using the intercept method. *Tectonophysics*, **267**, 91–119.

Launeau, P., Bouchez, J. L. & Benn, K. (1990). Shape preferred orientation of object populations; automatic analysis of digitized images. *Tectonophysics*, **180**, 201–11.

Launeau, P., Cruden, A. R. & Bouchez, J. L. (1994). Mineral recognition in digital images of rocks – a new approach using multichannel classification. *Canadian Mineralogist*, **32**, 919–33.

Lemelle, L., Simionovici, A., Truche, R. *et al.* (2004). A new nondestructive X-ray method for the determination of the 3D mineralogy at the micrometer scale. *American Mineralogist*, **89**, 547–53.

Lentz, R. C. F. & McSween, H. Y., Jr. (2000). Crystallization of the basaltic shergottites; insights from crystal size distribution (CSD) analysis of pyroxenes. *Meteoritics & Planetary Science*, **35**, 919–27.

Lewis, D. W. & McConchie, D. (1994a). *Analytical Sedimentology*. New York: Chapman & Hall.

Lewis, D. W. & McConchie, D. (1994b). *Practical Sedimentology*. New York: Chapman & Hall.

Lifshitz, I. M. & Slyozov, V. V. (1961). The kinetics of precipitation from supersaturated solid solutions. *Journal of Physics and Chemistry of Solids*, **19**, 35–50.

Lofgren, G. E. (1974). An experimental study of plagioclase crystal morphology: isothermal crystallization. *American Journal of Science*, **274**, 243–73.

Lofgren, G. E. & Donaldson, C. H. (1975). Curved branching crystals and differentiation in comb-layered rocks. *Contributions to Mineralogy and Petrology*, **49**, 309–19.

Maaloe, S., Tumyr, O. & James, D. (1989). Population density and zoning of olivine phenocrysts in tholeiites from Kauai, Hawaii. *Contributions to Mineralogy and Petrology*, **101**, 176–86.

Mainprice, D. & Nicolas, A. (1989). Development of shape and lattice preferred orientations – application to the seismic anisotropy of the lower crust. *Journal of Structural Geology*, **11**, 175–89.

Mandelbrot, B. B. (1982). *The Fractal Geometry of Nature*. San Francisco: W. H. Freeman.

Manga, M. (1998). Orientation distribution of microlites in obsidian. *Journal of Volcanology and Geothermal Research*, **86**, 107–15.

Mangan, M. T. (1990). Crystal size distribution and the determination of magma storage times: The 1959 eruption of Kilauea volcano, Hawaii. *Journal of Volcanology and Geothermal Research*, **44**, 295–302.

Mangan, M. T. & Cashman, K. V. (1996). The structure of basaltic scoria and reticulite and inferences for vesiculation, foam formation, and fragmentation in lava fountains. *Journal of Volcanology and Geothermal Research*, **73**, 1–18.

Mangan, M. T., Cashman, K. V. & Newman, S. (1993). Vesiculation of basaltic magma during eruption. *Geology*, **21**, 157–60.

Mardia, K. V. & Jupp, P. E. (2000). *Directional Statistics*. Chichester; New York: John Wiley & Sons.

Markov, I. V. (1995). *Crystal Growth for Beginners: Fundamentals of Nucleation, Crystal Growth, and Epitaxy*. Singapore: River Edge N.J.

Marqusee, J. A. & Ross, J. (1983). Kinetics of phase-transitions – theory of Ostwald ripening. *Journal of Chemical Physics*, **79**, 373–8.

Marschallinger, R. (1997). Automatic mineral classification in the macroscopic scale. *Computers & Geosciences*, **23**, 119–26.

Marschallinger, R. (1998a). Correction of geometric errors associated with the 3-D reconstruction of geological materials by precision serial lapping. *Mineralogical Magazine*, **62**, 783–92.

Marschallinger, R. (1998b). A method for three-dimensional reconstruction of macroscopic features in geological materials. *Computers & Geosciences*, **24**, 875–83.

Marschallinger, R. (2001). Three-dimensional reconstruction and visualization of geological materials with IDL – examples and source code. *Computers & Geosciences*, **27**, 419–26.

Marsh, B. D. (1988a). Crystal capture, sorting, and retention in convecting magma. *Geological Society of America Bulletin*, **100**, 1720–37.

Marsh, B. D. (1988b). Crystal size distribution (CSD) in rocks and the kinetics and dynamics of crystallization I. Theory. *Contributions to Mineralogy and Petrology*, **99**, 277–91.

Marsh, B. D. (1998). On the interpretation of crystal size distributions in magmatic systems. *Journal of Petrology*, **39**, 553–600.

Martin-Hernandez, F., Luneburg, C., Aubourg, C. & Jackson, M., eds., (2005). *Magnetic Fabric: Methods and Applications*. Geological Society Special Publication, 238. London: The Geological Society.

McBirney, A. R. & Hunter, R. H. (1995). The cumulate paradigm reconsidered. *Journal of Geology*, **103**, 114–22.

McBirney, A. R. & Nicolas, A. (1997). The Skaergaard layered series: Part II. Dynamic layering. *Journal of Petrology*, **38**, 569–80.

McConnell, J. (1975). Microstructures of minerals as petrogenetic indictors. *Annual Review of Earth and Planetary Sciences*, **3**, 129–55.

McKay, D. S., Gibson, E. K., Thomas-Keprta, K. L. *et al.* (1996). Search for past life on Mars: Possible relic biogenic activity in Martian meteorite ALH84001. *Science*, **273**, 924–30.

McKenzie, D. (1984). The generation and compaction of partially molten rock. *Journal of Petrology*, **25**, 713–65.

Medley, E. W. (2002). Estimating block size distributions of melanges and similar block-in-matrix rocks (bimrocks). In R. Hammah, W. Bawden, J. Curran & M. Telesnicki, eds., *North American Rock Mechanics Symposium*. Toronto, Canada: University of Toronto Press, pp. 509–606.

Mees, F., Swennen, R., Van Geet, M. & Jacobs, P., eds., (2003). *Applications of X-ray Computed Tomography in the Geosciences*. Geological Society Special Publication, 215. London: The Geological Society.

Meng, B. (1996). Determination and interpretation of fractal properties of the sandstone pore system. *Materials and Structures*, **29**, 195–205.

Merriam, D. F. (2004). The quantification of geology: from abacus to Pentium: A chronicle of people, places, and phenomena. *Earth-Science Reviews*, **67**, 55–89.

Meurer, W. P. & Boudreau, A. E. (1998). Compaction of igneous cumulates; Part II, Compaction and the development of igneous foliations. *Journal of Geology*, **106**, 293–304.

Middleton, G. V. (2000). *Data Analysis in the Earth Sciences Using MATLAB®*. Upper Saddle River, NJ: Prentice Hall.

Milligan, T. & Kranck, K. (1991). Electroresistance particle size analysers. In J. P. M. Syvitski, ed., *Principles, Methods, and Application of Particle Size Analysis*. New York: Cambridge University Press.

Miyazaki, K. (1996). A numerical simulation of textural evolution due to Ostwald ripening in metamorphic rocks: A case for small amount of volume of dispersed crystals. *Geochimica et Cosmochimica Acta*, **60**, 277–90.

Miyazaki, K. (2000). The case against Ostwald ripening of porphyroblasts: Discussion. *Canadian Mineralogist*, **38**, 1027–8.

Mock, A., Jerram, D. A. & Breitkreuz, C. (2003). Using quantitative textural analysis to understand the emplacement of shallow-level rhyolitic laccoliths – A case study from the Halle Volcanic Complex, Germany. *Journal of Petrology*, **44**, 833–49.

Mora, C. F. & Kwan, A. K. H. (2000). Sphericity, shape factor, and convexity measurement of coarse aggregate for concrete using digital image processing. *Cement and Concrete Research*, **30**, 351–8.

Morishita, R. (1998). Statistical properties of ideal rock textures: Relationship between crystal size distribution and spatial correlation in minerals. *Mathematical Geology*, **30**, 409–34.

Morishita, R. & Obata, M. (1995). A new statistical description of the spatial distribution of minerals in rocks. *Journal of Geology*, **103**, 232–40.

Morse, S. A. (1969). The Kiglapait layered intrusion, Labrador. *Geological Society of America, Memoir*, **112**, 204.

Morse, S. A. (1979). Kiglapait geochemistry, II. Petrography. *Journal of Petrology*, **20**, 591–624.

Muir, I. D. (1981). *The 4-Axis Universal Stage*. Chicago: Microscope Publications.

Mungall, J. & Su, S. (2005). Interfacial tension between sulfide and silicate liquids: Constraints on kinetics of sulfide liquation and sulfide migration through silicate rocks. *Earth and Planetary Science Letters*, **234**, 135–49.

Murthy, D. N. P., Xie, M. & Jiang, R. (2004). *Weibull Models*. Wiley Series in Probability and Statistics. Hoboken NJ: John Wiley and Sons.

Naslund, H. R. & McBirney, A. R. (1996). Mechanisms of formation of igneous layering. In R. G. Cawthorn, ed., *Layered Intrusions*. Amsterdam: Elsevier, pp. 1–44.

Nemchin, A. A., Giannini, L. M., Bodorkos, S. & Olivier, N. H. S. (2001). Ostwald ripening as a possible mechanism for zircon overgrowth formation during anatexis: theoretical constraints, a numerical model, and its application to pelitic migmatites of the Tickalra Metamorphics, northwestern Australia. *Geochimica et Cosmochimica Acta*, **65**, 2771–88.

Nesse, W. D. (1986). *Introduction to Optical Mineralogy*. New York: Oxford University Press.

Nicolas, A. (1992). Kinematics in magmatic rocks with special reference to gabbros. *Journal of Petrology*, **33**, 891–915.

Nicolas, A. & Ildefonse, B. (1996). Flow mechanism and viscosity in basaltic magma chambers. *Geophysical Research Letters*, **23**, 2013–16.

Orford, J. D. & Whalley, W. B. (1983). The use of the fractal dimension to quantify the morphology of irregular-shaped particles. *Sedimentology*, **30**, 655–68.

Ortoleva, P. J. (1994). *Geochemical Self-Organisation*. New York: Oxford University Press.

Pagel, M. (2000). *Cathodoluminescence in Geosciences*. Berlin: Springer.

Palmer, H. C. & MacDonald, W. D. (1999). Anisotropy of magnetic susceptibility in relation to source vents of ignimbrites: empirical observations. *Tectonophysics*, **307**, 207–18.

Palmer, H. C. & MacDonald, W. D. (2002). The Northeast Nevada Volcanic Field: Magnetic properties and source implications. *Journal of Geophysical Research-Solid Earth*, **107**, (B11), article number 2298.

Pan, Y. (2001). Inherited correlation in crystal size distribution. *Geology*, **29**, 227–30.

Pareschi, M., Pompilio, M. & Innocenti, F. (1990). Automated evaluation of spatial grain size distribution density from thin section images. *Computers & Geosciences*, **16**, 1067–84.

Park, H. H. & Yoon, D. N. (1985). Effect of dihedral angle on the morphology of grains in a matrix phase. *Metallurgical Transactions A-Physical Metallurgy and Materials Science*, **16**, 923–8.

Parsons, I., ed., (1987). *Origins of Igneous Layering*. NATO ASI series. Series C, Mathematical and Physical Sciences. Vol. 196. Dordrecht: D. Reidel Pub. Co., Boston/Norwell, MA, USA: Kluwer Academic Publishers.

Pearce, T. H. & Clark, A. H. (1989). Nomarski interference contrast observations of textural details in volcanic rocks. *Geology*, **17**, 757–9.

Pearce, T. H., Russell, J. K. & Wolfson, I. (1987). Laser-interference and Nomarski interference imaging of zoning profiles in plagioclase phenocrysts from the May 18, 1980, eruption of Mount St. Helens, Washington. *American Mineralogist*, **72**, 1131–43.

Perring, C. S., Barnes, S. J., Verrall, M. & Hill, R. E. T. (2004). Using automated digital image analysis to provide quantitative petrographic data on olivine-phyric basalts. *Computers & Geosciences*, **30**, 183–95.

Peterson, T. D. (1990). Petrology and genesis of natrocarbonatite. *Contributions to Mineralogy and Petrology*, **105**, 143–55.

Peterson, T. D. (1996). A refined technique for measuring crystal size distributions in thin section. *Contributions to Mineralogy and Petrology*, **124**, 395–405.

Petford, N., Davidson, G. & Miller, J. A. (2001). Investigation of the petrophysical properties of a porous sandstone sample using confocal scanning laser microscopy. *Petroleum Geoscience*, **7**, 99–105.

Petrik, I., Nabelek, P. I., Janak, M. & Plasienka, D. (2003). Conditions of formation and crystallization kinetics of highly oxidized pseudo tachylytes from the high Tatras (Slovakia). *Journal of Petrology*, **44**, 901–27.

Petruk, W. (1989). *Image Analysis in Earth Sciences*. Ottawa: Mineralogical Association of Canada.

Philpotts, A. R. & Dickson, L. D. (2000). The formation of plagioclase chains during convective transfer in basaltic magma. *Nature*, **406**, 59–61.

Philpotts, A. R. & Dickson, L. D. (2002). Millimeter-scale modal layering and the nature of the upper solidification zone in thick flood-basalt flows and other sheets of magma. *Journal of Structural Geology*, **24**, 1171–7.

Philpotts, A. R., Shi, J. Y. & Brustman, C. (1998). Role of plagioclase crystal chains in the differentiation of partly crystallized basaltic magma. *Nature*, **395**, 343–6.

Philpotts, A. R., Brustman, C. M., Shi, J. Y., Carlson, W. D. & Denison, C. (1999). Plagioclase-chain networks in slowly cooled basaltic magma. *American Mineralogist*, **84**, 1819–29.

Pickering, G., Bull, J. M. & Sanderson, D. J. (1995). Sampling power-law distributions. *Tectonophysics*, **248**, 1–20.

Pirard, E. (2004). Multispectral imaging of ore minerals in optical microscopy. *Mineralogical Magazine*, **68**, 323–33.

Polacci, M., Cashman, K. V. & Kauahikaua, J. P. (1999). Textural characterization of the pahoehoe-a'a transition in Hawaiian basalt. *Bulletin of Volcanology*, **60**, 595–609.

Poland, M. P., Fink, J. H. & Tauxe, L. (2004). Patterns of magma flow in segmented silicic dikes at Summer Coon volcano, Colorado: AMS and thin section analysis. *Earth and Planetary Science Letters*, **219**, 155–69.

Prince, C. M., Ehrlich, R. & Anguy, Y. (1995). Analysis of spatial order in sandstones. 2. Grain clusters, packing flaws, and the small-scale structure of sandstones. *Journal of Sedimentary Research Section A-Sedimentary Petrology and Processes*, **65**, 13–28.

Prior, D. J., Boyle, A. P., Brenker, F. *et al.* (1999). The application of electron backscatter diffraction and orientation contrast imaging in the SEM to textural problems in rocks. *American Mineralogist*, **84**, 1741–59.

Prior, D. J., Wheeler, J., Peruzzo, L., Spiess, R. & Storey, C. (2002). Some garnet microstructures: an illustration of the potential of orientation maps and misorientation analysis in microstructural studies. *Journal of Structural Geology*, **24**, 999–1011.

Proussevitch, A. A. & Sahagian, D. L. (2001). Recognition and separation of discrete objects within complex 3D voxelized structures. *Three-Dimensional Reconstruction, Modelling and Visualization of Geological Materials*, **27**, 441–54.

Pupin, J. P. (1980). Zircon and granite petrology. *Contributions to Mineralogy and Petrology*, **73**, 207–20.

Randle, V. & Caul, M. (1996). Representation of electron backscatter diffraction data. *Materials Science and Technology*, **12**, 844–50.

Randle, V. & Engler, O. (2000). *Introduction to Texture Analysis: Macrotexture, Microtexture and Orientation Mapping*. Reading, UK: Gordon and Breach Science Publishers.

Randolph, A. D. & Larson, M. A. (1971). *Theory of Particulate Processes*. New York: Academic Press.

Reed, S. J. B. (1996). *Electron Microprobe Analysis and Scanning Electron Microscopy in Geology*. Cambridge, New York: Cambridge University Press.

Reimann, C. & Filzmoser, P. (2000). Normal and lognormal data distribution in geochemistry; death of a myth; consequences for the statistical treatment of geochemical and environmental data. *Environmental Geology*, **39**, 1001–14.

Resmini, R. G. & Marsh, B. D. (1995). Steady-state volcanism, paleoeffusion rates, and magma system volume inferred from plagioclase crystal size distributions in mafic lavas; Dome Mountain, Nevada. *Journal of Volcanology and Geothermal Research*, **68**, 273–96.

Riegger, O. & van Vlack, L. (1960). Dihedral angle measurement. *Metallurgical Society of the American Institute of Metallurgical Engineers Transactions*, **218**, 933–5.

Robin, P. Y. F. (2002). Determination of fabric and strain ellipsoids from measured sectional ellipses – theory. *Journal of Structural Geology*, **24**, 531–44.

Rochette, P., Aubourg, C. & Perrin, M. (1999). Is this magnetic fabric normal? A review and case studies in volcanic formations. *Tectonophysics*, **307**, 219–34.

Rochette, P., Jackson, M. & Aubourg, C. (1992). Rock magnetism and the interpretation of anisotropy of magnetic-susceptibility. *Reviews of Geophysics*, **30**, 209–26.

Rodriguez-Navarro, A. B. & Romanek, C. S. (2002). Mineral fabrics analysis using a low-cost universal stage for X-ray diffractometry. *European Journal of Mineralogy*, **14**, 987–92.

Rogers, C. D. F., Dijkstra, T. A. & Smalley, I. J. (1994). Particle packing from an earth-science viewpoint. *Earth-Science Reviews*, **36**, 59–82.

Rombouts, L. (1995). Sampling and statistical evaluation of diamond deposits. *Journal of Geochemical Exploration*, **53**, 351–67.

Rosenberg, C. L. & Handy, M. R. (2005). Experimental deformation of partially melted granite revisited: implications for the continental crust. *Journal of Metamorphic Geology*, **23**, 19–28.

Ross, B. J., Fueten, F. & Yashkir, D. Y. (2001). Automatic mineral identification using genetic programming. *Machine Vision and Applications*, **13**, 61–9.

Royet, J.-P. (1991). Stereology: A method for analysing images. *Progress in Neurobiology*, **37**, 433–74.

Rubin, A. E. (2000). Petrologic, geochemical and experimental constraints on models of chondrule formation. *Earth-Science Reviews*, **50**, 3–27.

Rubin, A. E. & Grossman, J. N. (1987). Size-frequency-distributions of EH3 chondrules. *Meteoritics*, **22**, 237–51.

Rudashevsky, N. S., Burakov, B. E., Lupal, S. D., Thalhammer, O. A. R. & Sainieidukat, B. (1995). Liberation of accessory minerals from various rock types by electric-pulse disintegration-method and application. *Transactions of the Institution of Mining and Metallurgy Section C-Mineral Processing and Extractive Metallurgy*, **104**, C25–C29.

Russ, J. C. (1986). *Practical Stereology*. New York: Plenum Press.

Russ, J. C. (1999). *The Image Processing Handbook*. Boca Raton, Florida, USA: CRC Press.

Rust, A. C. & Cashman, K. V. (2004). Permeability of vesicular silicic magma: inertial and hysteresis effects. *Earth and Planetary Science Letters*, **228**, 93–107.

Rust, A. C., Manga, M. & Cashman, K. V. (2003). Determining flow type, shear rate and shear stress in magmas from bubble shapes and orientations. *Journal of Volcanology and Geothermal Research*, **122**, 111–32.

Sahagian, D. L. & Maus, J. E. (1994). Basalt vesicularity as a measure of atmospheric-pressure and palaeoelevation. *Nature*, **372**, 449–51.

Sahagian, D. L. & Proussevitch, A. A. (1998). 3D particle size distributions from 2D observations; stereology for natural applications. *Journal of Volcanology and Geothermal Research*, **84**, 173–96.

Sahagian, D. L., Proussevitch, A. A. & Carlson, W. D. (2002a). Analysis of vesicular basalts and lava emplacement processes for application as a paleobarometer/paleoaltimeter. *Journal of Geology*, **110**, 671–85.

Sahagian, D. L., Proussevitch, A. A. & Carlson, W. L. (2002b). Timing of Colorado Plateau uplift: Initial constraints from vesicular basalt-derived paleoelevations. *Geology*, **30**, 807–10.

Saiki, K. (1997). Morphology and simulation of solid state rounding process. *Geophysical Research Letters*, **24**, 1519–22.

Saiki, K., Laporte, D., Vielzeuf, D., Nakashima, S. & Boivin, P. (2003). Morphological analysis of olivine grains annealed in an iron-nickel matrix: Experimental constraints on the origin of pallasites and on the thermal history of their parent bodies. *Meteoritics & Planetary Science*, **38**, 427–44.

Saint-Blanquat, M. & Tikoff, B. (1997). Development of magmatic to solid-state fabrics during syntectonic emplacement of the Mono Creek granite, Sierra Nevada Batholith. In J. L. Bouchez, D. H. W. Hutton, & W. E. Stephens, eds., *Granite: From Segregation of Melt to Emplacement Fabrics*. Dordrecht: Kluwer Academic Publishers, pp. 231–52.

Saltikov, S. A. (1967). The determination of the size distribution of particles in an opaque material from a measurement of the size distributions of their sections. In H. Elias, ed., *Proceedings of the Second International Congress for Stereology*. Berlin: Springer-Verlag, pp. 163–73.

Saltzer, R. L., Chatterjee, N. & Grove, T. L. (2001). The spatial distribution of garnets and pyroxenes in mantle peridotites: Pressure-temperature history of peridotites from the Kaapvaal craton. *Journal of Petrology*, **42**, 2215–29.

Sato, H. (1995). Textural difference between pahoehoe and a'a lavas of Izu-Oshima volcano, Japan; an experimental study on population density of plagioclase. *Models of Magmatic Processes and Volcanic Eruptions*, **66**, 101–13.

Schafer, F. & Foley, S. F. (2002). The effect of crystal orientation on the wetting behaviour of silicate melts on the surfaces of spinel peridotite minerals. *Contributions to Mineralogy and Petrology*, **143**, 254–61.

Schafer, W. (2002). Neutron diffraction applied to geological texture and stress analysis. *European Journal of Mineralogy*, **14**, 263–89.

Schmid, S. M., Casey, M. & Starkey, J. (1981). An illustration of the advantages of a complete texture analysis described by the orientation distribution function (ODF) using quartz pole figure data. *Tectonophysics*, **78**, 101–17.

Schwindinger, K. R. (1999). Particle dynamics and aggregation of crystals in a magma chamber with application to Kilauea Iki olivines. *Journal of Volcanology and Geothermal Research*, **88**, 209–38.

Schwindinger, K. R. & Anderson, A. T. (1989). Synneusis of Kilauea Iki olivines. *Contributions to Mineralogy and Petrology*, **103**, 187–98.

Scott, R. G. & Benn, K. (2001). Peak-ring rim collapse accommodated by impact melt-filled transfer faults, Sudbury impact structure, Canada. *Geology*, **29**, 747–50.

Sempels, J.-M. (1978). Evidence for constant habit development of plagioclase crystals from igneous rocks. *Canadian Mineralogist*, **16**, 257–63.

Shelley, D. (1985). Determining paleo-flow directions from groundmass fabrics in the Lyttelton radial dykes, New Zealand. *Journal of Volcanology and Geothermal Research*, **25**, 69–79.

Shin, H., Lindquist, W. B., Sahagian, D. L. & Song, S. R. (2005). Analysis of the vesicular structure of basalts. *Computers & Geosciences*, **31**, 473–87.

Shore, M. & Fowler, A. D. (1999). The origin of spinifex texture in komatiites. *Nature*, **397**, 691–4.

Smit, T. H., Schneider, E. & Odgaard, A. (1998). Star length distribution: a volume-based concept for the characterization of structural anisotropy. *Journal of Microscopy-Oxford*, **191**, 249–57.

Smith, C. S. (1948). Grains, phases and interfaces: an interpretation of microstructure. *Transactions of the Metallurgical Society of the AIME*, **175**, 15–51.

Smith, C. S. (1964). Some elementary principles of polycrystalline microstructure. *Metallurgical Review*, **9**, 1–48.

Smith, J. V. (2002). Structural analysis of flow-related textures in lavas. *Earth-Science Reviews*, **57**, 279–97.

Song, S. R., Jones, K. W., Lindquist, W. B., Dowd, B. A. & Sahagian, D. L. (2001). Synchrotron X-ray computed microtomography: studies on vesiculated basaltic rocks. *Bulletin of Volcanology*, **63**, 252–63.

Sparks, R. S. J. (1978). Dynamics of bubble formation and growth in magmas – review and analysis. *Journal of Volcanology and Geothermal Research*, **3**, 1–37.

Sparks, R. S. J. (1997). Causes and consequences of pressurisation in lava dome eruptions. *Earth and Planetary Science Letters*, **150**, 177–89.

Spiess, R., Peruzzo, L., Prior, D. J. & Wheeler, J. (2001). Development of garnet porphyroblasts by multiple nucleation, coalescence and boundary misorientation-driven rotations. *Journal of Metamorphic Geology*, **19**, 269–90.

Stamatelopoulou-Seymour, K., Vlassopoulos, D., Pearce, T. H. & Rice, C. (1990). The record of magma chamber processes in plagioclase phenocrysts at Thera volcano, Aegean Volcanic Arc, Greece. *Contributions to Mineralogy and Petrology*, **104**, 73–84.

Stipp, M. & Tullis, J. (2003). The recrystallized grain size piezometer for quartz. *Geophysical Research Letters*, **30**, article number 2088.

Sturm, R. (2004). Imaging of growth banding of minerals using 2-stage sectioning: application to accessory zircon. *Micron*, **35**, 681–4.

Sunagawa, I. (1987a). *Morphology of Crystals*. Dordrecht: D. Reidel.

Sunagawa, I. (1987b). Morphology of minerals. In I. Sunagawa, ed., *Morphology of Crystals*. Dordrecht: D. Reidel, pp. 509–88.

Suteanu, C. & Kruhl, J. H. (2002). Investigation of heterogeneous scaling intervals exemplified by sutured quartz grain boundaries. *Fractals-Complex Geometry Patterns and Scaling in Nature and Society*, **10**, 435–49.

Suteanu, C., Zugravescu, D. & Munteanu, F. (2000). Fractal approach of structuring by fragmentation. *Pure and Applied Geophysics*, **157**, 539–57.

Sutton, A. P. & Balluffi, R. W. (1996). *Interfaces in Crystalline Materials*. Oxford, UK: Oxford Science Publications.

Swan, A. R. H. & Sandilands, M. (1995). *Introduction to Geological Data Analysis*. Cambridge, MA, USA: Blackwell Science.

Syvitski, J. P. M. (1991). *Principles, Methods, and Application of Particle Size Analysis*. New York: Cambridge University Press.

Syvitski, J. P. M., Asprey, K. & Clattenberg, D. (1991). Principles, design, and calibration of settling tubes. In J. P. M. Syvitski, ed., *Principles, Methods, and Application of Particle Size Analysis*. New York: Cambridge University Press, pp. 45–63.

Takahashi, M. & Nagahama, H. (2000). Fractal grain boundary migration. *Fractals*, **8**, 189–94.

Tarling, D. H. & Hrouda, F. (1993). *The Magnetic Anisotropy of Rocks*. London: Chapman & Hall.

Tarquini, S. & Armienti, P. (2001). Film color scanner as a new and cheap tool for image analysis in petrology. *Image Analysis and Stereology*, **20** (Suppl. 1), 567–72.

Taylor, L. (2000). Diamonds and their mineral inclusions, and what they tell us: A detailed pull-part of a diamondiferous eclogite. *International Geology Review*, **42**, 959–83.

Taylor, L. A., Nazarov, M. A., Shearer, C. K. *et al.* (2002). Martian meteorite Dhofar 019: A new shergottite. *Meteoritics & Planetary Science*, **37**, 1107–28.

Thomas, M. C., Wiltshire, R. J. & Williams, A. T. (1995). The use of Fourier descriptors in the classification of particle-shape. *Sedimentology*, **42**, 635–45.

Thomas-Keprta, K. L., Bazylinski, D. A., Kirschvink, J. L. *et al.* (2000). Elongated prismatic magnetite crystals in ALH84001 carbonate globules: Potential Martian magnetofossils. *Geochimica et Cosmochimica Acta*, **64**, 4049–81.

Thompson, S., Fueten, F. & Bockus, D. (2001). Mineral identification using artificial neural networks and the rotating polarizer stage. *Computers & Geosciences*, **27**, 1081–9.

Titkov, S. V., Saparin, G. V. & Obyden, S. K. (2002). Evolution of growth sectors in natural diamond crystals as revealed by cathodoluminescence topography. *Geology of Ore Deposits*, **44**, 350–60.

Toramaru, A. (1989). Vesiculation process and bubble size distributions in ascending magmas with constant velocities. *Journal of Geophysical Research, B, Solid Earth and Planets*, **94**, 17523–42.

Toramaru, A. (1990). Measurement of bubble-size distributions in vesiculated rocks with implications for quantitative estimation of eruption processes. *Journal of Volcanology and Geothermal Research*, **43**, 71–90.

Treiman, A. H. (2003). Submicron magnetite grains and carbon compounds in Martian meteorite ALH84001: Inorganic, abiotic formation by shock and thermal metamorphism. *Astrobiology*, **3**, 369–92.

Trindade, R. I. F., Bouchez, J. L., Bolle, O. *et al.* (2001). Secondary fabrics revealed by remanence anisotropy: methodological study and examples from plutonic rocks. *Geophysical Journal International*, **147**, 310–18.

Tuffen, H. (1998). L'origine des cristaux dans le chambre magmatique de Santorin (Grèce). Clermont-Ferrand, France: Université Blaise-Pascal.

Turcotte, D. L. (1992). *Fractals and Chaos in Geology and Geophysics*. Cambridge, New York: Cambridge University Press.

Turner, S., George, R., Jerram, D. A., Carpenter, N. & Hawkesworth, C. (2003). Case studies of plagioclase growth and residence times in island arc lavas from Tonga and the Lesser Antilles, and a model to reconcile discordant age information. *Earth and Planetary Science Letters*, **214**, 279–94.

Underwood, E. E. (1970). *Quantitative Stereology*. Reading, MA: Addison-Wesley.

van den Berg, E. H., Meesters, A. G. C. A., Kenter, J. A. M. & Schlager, W. (2002). Automated separation of touching grains in digital images of thin sections*1. *Computers & Geosciences*, **28**, 179–90.

Vance, J. A. (1969). On synneusis. *Contributions to Mineralogy and Petrology*, **24**, 7–29.

Vavra, G. (1993). A guide to quantitative morphology of accessory zircon. *Chemical Geology*, **110**, 15–28.

Ventura, G. (2001). The strain path and emplacement mechanism of lava flows: an example from Salina (southern Tyrrhenian Sea, Italy). *Earth and Planetary Science Letters*, **188**, 229–40.

Ventura, G., DeRosa, R., Colletta, E. & Mazzuoli, R. (1996). Deformation patterns in a high-viscosity lava flow inferred from the crystal preferred orientation and imbrication structures: An example from Salina (Aeolian Islands, southern Tyrrhenian Sea, Italy). *Bulletin of Volcanology*, **57**, 555–62.

Vernon, R. (1970). Comparative grain boundary studies of some basic and ultrabasic granulites, nodules and cumulates. *Scottish Journal of Geology*, **6**, 337–51.

Vernon, R. H. (1968). Microstructures of high-grade metamorphic rocks at Broken Hill, Australia. *Journal of Petrology*, **9**, 1–22.

Vernon, R. H. (1986). K-feldspar megacrysts in granites – phenocrysts not porphyroblasts. *Earth-Science Reviews*, **23**, 1–63.

Vernon, R. H. (2004). *A Practical Guide to Rock Microstructure*. Cambridge: Cambridge University Press.

Verrecchia, E. P. (2003). Foreword: image analysis and morphometry of geological objects. *Mathematical Geology*, **35**, 759–62.

Vigneresse, J. L., Barbey, P. & Cuney, M. (1996). Rheological transitions during partial melting and crystallization with application to felsic magma segregation and transfer. *Journal of Petrology*, **37**, 1579–1600.

Voorhees, P. W. (1992). Ostwald ripening of two-phase mixtures. *Annual Review of Materials Science*, **22**, 197–215.

Wada, Y. (1992). Magma flow directions inferred from preferred orientations of phenocryst in a composite feeder dyke, Miyake-Jima, Japan. *Journal of Volcanology and Geothermal Research*, **49**, 119–26.

Waff, H. S. & Bulau, J. R. (1979). Equilibrium fluid distribution in an ultramafic partial melt under hydrostatic stress conditions. *Journal of Geophysical Research*, **84**, 6109–14.

Wager, L. R. (1961). A note on the origin of ophitic texture in the chilled olivine gabbro of the Skaergaard intrusion. *Geological Magazine*, **98**, 353–66.

Wager, L. R. & Brown, G. M. (1968). *Layered Igneous Rocks*. Edinburgh; London: Oliver & Boyd.

Waters, C. & Boudreau, A. E. (1996). A re-evaluation of crystal size distribution in chromite cumulates. *American Mineralogist*, **81**, 1452–9.

Watson, E. B. & Brenan, J. M. (1987). Fluids in the lithosphere.1. experimentally-determined wetting characteristics of CO_2-H_2O fluids and their implications for fluid transport, host-rock physical-properties, and fluid inclusion formation. *Earth and Planetary Science Letters*, **85**, 497–515.

Wegner, M. & Christie, J. (1985). General chemical etchants for microstructures and defects in silicates. *Physics and Chemistry of Minerals*, **12**, 90–2.

Wenk, H. R. (2002). Texture and anisotropy. In *Plastic Deformation of Minerals and Rocks*. Reviews in Mineralogy & Geochemistry, 51. Washington DC: Mineralogical Society of America, pp. 291–329.

Wenk, H. R. (1985). *Preferred Orientation in Deformed Metals and Rocks: An Introduction to Modern Texture Analysis*. Orlando FL: Academic Press.

Wenk, H. R. & Grigull, S. (2003). Synchrotron texture analysis with area detectors. *Journal of Applied Crystallography*, **36**, 1040–9.

Wenk, H. R. & Van Houtte, P. (2004). Texture and anisotropy. *Reports on Progress in Physics*, **67**, 1367–428.

Whitham, A. & Sparks, R. (1986). Pumice. *Bulletin of Volcanology*, **48**, 209–23.

Wilson, B., Dewers, T., Ze'ev, R. & Brune, J. (2005). Particle size and energetics of gouge from earthquake rupture zones. *Nature*, **434**, 749–52.

Wright, I. C., Gamble, J. A. & Shane, P. A. R. (2003). Submarine silicic volcanism of the Healy caldera, southern Kermadec arc (SW Pacific): I – volcanology and eruption mechanisms. *Bulletin of Volcanology*, **65**, 15–29.

Wulff, G. (1901). Zur frage der Geschwindigkeit des Wachstums und der Auflosumg der Krystallflachen. *Zeitschift für Kristallographie und Mineralogie*, **34**, 449–530.

Xie, Y. X., Wenk, H. R. & Matthies, S. (2003). Plagioclase preferred orientation by TOF neutron diffraction and SEM-EBSD. *Tectonophysics*, **370**, 269–86.

Yaouancq, G. & MacLeod, C. J. (2000). Petrofabric investigation of gabbros from the Oman ophiolite: comparison between AMS and rock fabric. *Marine Geophysical Researches*, **21**, 289–305.

Zeh, A. (2004). Crystal size distribution (CSD) and textural evolution of accessory apatite, titanite and allanite during four stages of metamorphism: an example from the Moine supergroup, Scotland. *Journal of Petrology*, **45**, 2101–32.

Zellmer, G., Turner, S. & Hawkesworth, C. (2000). Timescales of destructive plate margin magmatism; new insights from Santorini, Aegean volcanic arc. *Earth and Planetary Science Letters*, **174**, 265–81.

Zhou, Y., Starkey, J. & Mansinha, L. (2004a). Identification of mineral grains in a petrographic thin section using phi- and max-images. *Mathematical Geology*, **36**, 781–801.

Zhou, Y., Starkey, J. & Mansinha, L. (2004b). Segmentation of petrographic images by integrating edge detection and region growing. *Computers & Geosciences*, **30**, 817–31.

Zieg, M. J. & Marsh, B. D. (2002). Crystal size distributions and scaling laws in the quantification of igneous textures. *Journal of Petrology*, **43**, 85–101.

Index